Lecture Notes in Computer Science 7720

Commenced Publication in 1973
Founding and Former Series Editors:
Gerhard Goos, Juris Hartmanis, and Jan van Leeuwen

T0238966

Editorial Board

Abdelkader Hameurlain   Josef Küng
Roland Wagner (Eds.)

Transactions on Large-Scale Data- and Knowledge-Centered Systems VII

Springer

Editors-in-Chief

Abdelkader Hameurlain
Paul Sabatier University, IRIT
118 route de Narbonne, 31062 Toulouse Cedex, France
E-mail: hameur@irit.fr

Josef Küng
Roland Wagner
University of Linz, FAW
Altenbergerstraße 69, 4040 Linz, Austria
E-mail: {jkueng,rrwagner}@faw.at

ISSN 0302-9743 (LNCS)   e-ISSN 1611-3349 (LNCS)
ISSN 1869-1994 (TLDKS)
ISBN 978-3-642-35331-4   e-ISBN 978-3-642-35332-1
DOI 10.1007/978-3-642-35332-1
Springer Heidelberg Dordrecht London New York

Library of Congress Control Number: 2012952682

CR Subject Classification (1998): H.2.4, H.2.8, I.2, E.2, I.6.5, H.3.1, D.3.2-3, G.3, E.1

Typesetting: Camera-ready by author, data conversion by Scientific Publishing Services, Chennai, India

Printed on acid-free paper

Springer is part of Springer Science+Business Media (www.springer.com)

Preface

This volume is the second so-called regular volume of our journal. We find here the result of the review process covering all the submissions that have been sent directly to the journal administration since our last regular volume in 2012. Recognized scientists checked the quality and gave helpful feedback to the authors – many thanks to them from the editorial board.

This review process resulted in the selection of five contributions. Their range is from data management via data streams to service oriented computing, and from an abstract algebraic framework via RDF and ontologies to a conceptual model framework.

In the first contribution, *'RFID Data Management and Analysis via Tensor Calculus'*, Roberto De Virgilio and Franco Milicchio meet the challenge of managing the increasing amount of RFID data in supply chains. They introduce a novel algebraic framework for modeling a supply chain and for efficiently performing analysis. The main advantage of this approach is its theoretical soundness. Tensor calculus provides the background for both modeling and querying.

Then, Abhirup Chakraborty and Ajit Singh address the issue of processing exact sliding window joins between data streams in a memory-limited environment having burstiness in stream arrivals. *'Processing Exact Results for Windowed Stream Joins in a Memory-Limited System: A Disk-Based Approach'* provides a framework and proposes an algorithm to solve that problem. Disk storage and a smart I/O strategy are used to meet the goals.

In the third article, we switch to semantic issues. In *'Reducing the Semantic Heterogeneity of Unstructured P2P Systems: A Contribution Based on a Dissemination Protocol'* Thomas Cerqueus, Sylvie Cazalens and Philippe Lamarre address the situation that arises when, in a peer-to-peer system, different peers are using different ontologies to annotate their data. First they propose a set of measures to characterize different facets of heterogeneity. Then they introduce a gossip-based protocol that allows some of them to be reduced.

'Towards a Scalable Semantic Provenance Management System', written by Mohamed Amin Sakka and Bruno Defude, presents a new provenance management system. It allows provenance sources to be imported and enriched semantically to obtain a high-level representation of provenance. The power of semantic web technologies is used for heterogeneous, multiple source and decentralized provenance integration. So rich answers about the whole document life cycle can be provided and trustworthiness can be increased.

Finally, Colin Atkinson, Philipp Bostan and Dirk Draheim concentrate on service-oriented distributed systems. In *'A Unified Conceptual Framework for Service-Oriented Computing: Aligning Models of Architecture and Utilization'* they support various concepts and models and thereby make it possible to customize and simplify each client developer's view as well as the way in which service providers develop and maintain their services. The paper provides a unified conceptual framework for describing the components of service oriented computing systems, in particular its foundations and concepts, and a small example to show how the models are applied in practice.

Last but not least we would like to thank Gabriela Wagner for supporting us with the organization and we hope that you enjoy this TLDKS volume.

October 2012   Abdelkader Hameurlein
                                                          Josef Küng
                                                          Roland Wagner

Editorial Board

Table of Contents

RFID Data Management and Analysis via Tensor Calculus*

Roberto De Virgilio and Franco Milicchio

Dipartimento di Informatica e Automazione
Universitá Roma Tre, Rome, Italy
{dvr,milicchio}@dia.uniroma3.it

Abstract. Traditional Radio-Frequency IDentication (RFID) applications have been focused on replacing bar codes in supply chain management. The importance of such new resource soared in recent years, mainly due to the retailers' need of governing supply chains. However, due to the massive amount of RFID-related information in supply chain management, attaining satisfactory performances in analyzing such data sets is a challenging issue. Popular approaches provide hard-coded solutions, with high consumption of resources; moreover, these exhibit very inadequate adaptability when dealing with multidimensional queries, at various levels of granularity and complexity.

In this paper we propose a novel model for supply chain management, aiming at generality, correctness, and simplicity. Such model is based on the first principles of multilinear algebra, specifically, of *tensorial calculus*.

Leveraging our abstract algebraic framework, we envision a system allowing both quick decentralized on-line item discovery and centralized off-line massive business logic analysis, according to needs and requirements of supply chain actors. Being our computations based on vectorial calculus, we are able to exploit the underlying hardware processors, achieving a huge performance boost, as the experimental results show. Moreover, by storing only the needed data, and benefiting from linear properties, we are able to carry out the required computations even in high memory constrained environments, such as on mobile devices, and in parallel and distributed technologies by subdividing our tensor objects into sub-blocks, and processing them independently.

1 Introduction

A supply chain is a complex system for transferring products or services from a supplier to a final customer. Aiming at improving the exchange of information, retailers are investing in new technologies for managing supply chains: to this aim, RFID, the Radio-Frequency Identification, is a recent potential wireless technology, which enables a direct link of a product with a virtual one, within

---

* This work has been partially funded by the Italian Ministry of Research, grant number RBIP06BZW8, FIRB project "Advanced tracking system in intermodal freight transportation".

A. Hameurlain et al. (Eds.): TLDKS VII, LNCS 7720, pp. 1–30, 2012.

information systems [1, 2]. Generally, RFID tags attached to products, store an identification code named EPC—which stands for Electronic Product Code [3], a coding scheme for RFID tags, aimed at uniquely identify them—used as a key to retrieve relevant properties of an object from a database, usually within a networked infrastructure.

Usually, an RFID application generates a set of tuples, usually called *raw data*, of the form of a triple (e, l, t), where e is an EPC, l represents the location where an RFID reader has scanned the object e, and t is the time when the reading took place; while these are fundamental properties, others can be retrieved as well, *e.g.*, temperature, pressure, and humidity. Tags may have multiple subsequent readings at the same location, potentially generating vast amounts of raw data. Consequently, a simple data cleaning technique consists in converting raw data in *stay records*, *i.e.*, tuples structured as (e, l, t_i, t_o), where t_i and t_o are the time when an object enters or leaves a location l, respectively.

In this manuscript, we address the challenging problem of efficiently managing the tera-scale amount of data per day, generated by RFID applications (cf. [4], [5], [6], and [7]), focusing on stay records as the basic block to store RFID data. We advocate the use of a model based on the first principles of *tensorial algebra*, achieving a great deal of flexibility, while maintaining simplicity and correctness.

Recently, two models have been proposed: a *herd* and a *single* model. The former [6] proposes a warehousing model observing that, usually, products move together in large groups at the beginning, and continue in small herds towards the end of the chain: this allows the aggregation and reduces significantly the size of the database. On the contrary, the latter [7] focuses on the movement of single, non grouped tags, defining query templates and path encoding schemes to process tracking and path oriented queries efficiently.

Both of these approaches present limited flexibility dealing multidimensional queries at varying levels of granularity and complexity. To exploit the compression mechanism, such proposals have to fix the dimensions of analysis in advance and implement *ad-hoc* data structures to be maintained. It is not known in advance whether objects are sorted and grouped, and therefore, it is problematic to support different kinds of high-level queries efficiently.

In this paper we propose a novel approach based on multilinear algebra. Matrix operations are invaluable tools in several fields, from engineering and design, graph theory, or networking; in particular streams and co-evolving sequences can also be envisioned as matrices: each data source, a sensor, may correspond to a row, while time-ticks to a column (cf. [8]).

Standard matrix approaches focus on a standard two-dimension space, while we extend the applicability of such techniques with the more general definition of *tensors*, a generalization of linear forms, which are usually represented by matrices. We may therefore take advantage of the vast literature, both theoretic and applied, regarding tensor calculus. Computational algebra is employed in critical applications, such as engineering and physics, and several libraries have been developed with two major goals: *efficiency* and *accuracy*.

Contribution. Leveraging such background, this paper proposes a general model of supply chains, mirrored with a formal tensor representation and endowed with specific operators, allowing both quick decentralized on-line processing, and centralized off-line massive business logic analysis, according to needs and requirements of supply chain actors. Our model and operations, inherited by linear algebra and tensor calculus, is therefore theoretically sound, and its implementation benefits from numerical libraries developed in the past, exploiting the underlying hardware, optimized for mathematical operations. Additionally, due to the properties of our tensorial model, we are able to attain two significative features: the possibility of conducting computations in memory-constrained environments such as on mobile devices, and exploiting modern parallel and distributed technologies, owing to the possibility for matrices, and therefore tensors, to be dissected into several chunks, and processed independently (*i.e.*, also on-the-fly).

Outline. Our manuscript is organized as follows. In section 2 we will briefly recall the available literature, while Section 3 will be devoted to the introduction of tensors and their associated operations. The general supply chain model, accompanied by a formal tensorial representation is supplied in Section 4, subsequently put into practice in Sections 5 and 6, where we provide the reader a method of analyzing RFID data within our framework. We benchmark our approach, with asymptotic complexity outlined in Section 7, with several test beds, and supply the results in Section 8.

2   Related Work

Real-life scenarios are affected by great inconsistencies and errors in data, and face the problem of dealing with huge amounts of data, generated on a daily basis. To this aim, knowledge representation techniques focus on operating deep analysis in these systems. There exists two main approaches to the management of RFID data: on one hand, we may process data streams at run-time [9–12], while on the other hand, processing is performed off-line, once RFID data are aggregated, compressed, and stored [6, 7].

If we consider RFID data as a stream, the main issues are event processing and data cleaning. Wang et al. have proposed a conceptualization of RFID events based on an extension of the ER model [12]. Bai et al. have studied the limitations of using SQL to detect temporal events and have presented an SQL-like language to query such events in an efficient way [9]. Inaccuracies of RFID tags readings cause irregularities in the processing of data: therefore cleaning techniques need to be applied to RFID data streams as proposed in [10].

An alternative approach consists in warehousing RFID data, and performing multidimensional analyses on the warehouse. The focus here is on data compression techniques and on storage models, with the goal of achieving a more expressive and effective representation of RFID data. A straightforward method is to provide a support to path queries—*e.g.*, find the average time for products to go from factories to stores in Seattle—by collecting RFID tag movements along the supply chain. Usually, the tag identifier, the location and the time of each RFID reading is gathered and stored in a huge relational table. Gonzalez et al. [6] have presented a new storage model, based on a data warehousing approach, in which items moving together are grouped and data analysis is preformed in a multidimensional fashion, as it happens in a typical data warehouse. Finally, Lee et al. proposed an effective path encoding approach, aimed at representing the data flow abstracting movements of products [7]. With this approach, in each path, a prime number is assigned to each node, and a path is encoded as the product of the number associated with nodes. Mathematical properties of prime numbers guarantee the unambiguous access to paths.

As highlighted in the Introduction, a prime limitation of the majority of these approaches rely in the requirement to fix the dimensions of analysis in advance, so that *ad-hoc* data structures can be efficiently maintained. It follows that, in many application scenarios, the compression loses its effectiveness, and the size of tables is not significantly reduced. Moreover, most of the approaches presented in this section, present a limited flexibility when multidimensional queries, at varying levels of granularity and complexity, need to be performed.

3   Preliminary Issues

The following paragraphs will be devoted to a brief recall of tensor representations, which is preliminary to the introduction of our model in the ensuing sections.

We take for granted the elementary algebra notions, and their respective notations, of *groups*, *vector spaces*, *matrix rings*, and refer the reader to [13] for a brief review.

3.1   Tensors

Tensors arise as a natural extension of linear forms. Linear forms on a vector space $\mathbb{V}$ defined over a field $\mathbb{F}$, and belonging to its *dual space* $\mathbb{V}^*$, are maps $\phi : \mathbb{V} \to \mathbb{F}$ that associate to each pair $\phi \in \mathbb{V}^*$, and $v \in \mathbb{V}$, an element $e \in \mathbb{F}$. This is usually denoted by the *pairing* $\langle \phi, v \rangle = e$, or by definition, with a *functional* notation, *i.e.*, $\phi(v) = e$. Such mapping exhibits the linearity property, *i.e.*,

$$\langle \phi, \alpha\, v + \beta w \rangle = \alpha \langle \phi, v \rangle + \beta \langle \phi, w \rangle\,.$$

Generalizing the concept of linearity, we hence may introduce the following:

Definition 1 (Tensor). *A* tensor *is a multilinear form ϕ on a vector space $\mathbb{V}$, i.e., a mapping*

$$\phi : \underbrace{\mathbb{V} \times \ldots \times \mathbb{V}}_{k-\text{times}} \longrightarrow \mathbb{F}\,, \tag{1}$$

the form ϕ assumes also the name of tensor of rank k *(or rank-k tensor).*

The property of *multilinearity* mirrors the previous definition of linearity: ϕ is, in fact, linear in each of its k domains.

Matrices are the prominent descriptive form of linear operators, and analogously, tensors may be represented with elements of the *matrix ring*. Therefore, a rank-k tensor may be represented by $M \in \mathbb{M}_{i_1 i_2 \dots i_k}(\mathbb{F})$, that is a k-dimensional matrix with elements in $\mathbb{F}$.

A comprehensive description of tensors and their associated properties is beyond the scope of this manuscript; for a thorough review, see [14, 15], and [16].

3.2 Operations

Vectors and forms, both belonging to a *vector space*, provide different operations: from basic sum and product of a vector (form) with a scalar to more complex operations that may be defined on vector spaces. In particular, let us consider the following:

Definition 2 (Hadamard Product). *The* Hadamard product *of two vectors* $v, w \in \mathbb{V}$ *is the* entry-wise *product of their components, i.e.,*

$$v \circ w = r, \quad such\ that \quad r_i := v_i w_i. \tag{2}$$

To clarify the above definition, we provide the reader with the following example:

Example 1. Let us consider two vectors $v, w \in \mathbb{R}^3$, with $v = \begin{pmatrix} 1 & 4 & 0 \end{pmatrix}^t$, and $w = \begin{pmatrix} 2 & 3 & 9 \end{pmatrix}^t$, with $(\ )^t$ denoting the *transposed* vector. The Hadamard product will be

$$v \circ w = \begin{pmatrix} 1 \cdot 2 & 4 \cdot 3 & 0 \cdot 9 \end{pmatrix}^t = \begin{pmatrix} 2 & 12 & 0 \end{pmatrix}^t.$$

Since a tensor of rank k will henceforth be denoted by a multidimensional matrix, we shall simplify notations by a variant of *Einstein's notation*, or *summation convention*. Such variant will be employed wherever a summation occurs: where we encounter two identical indices in a given expression, a summation on that index is implicitly defined. This convention will be promptly clarified.

Example 2 (Einstein's Notation). Let us consider for example, a notable form, the *inner product* $(\cdot)$ between two vectors, known also as *scalar product*. If we consider a standard cartesian metrics, *i.e.*, an orthonormal basis [13], the inner product notation can be written leveraging the summation notation described above:

$$v \cdot w := \sum_j v_j w_j =: v_j w_j.$$

We mention in passing that an inner product arises in defining a particular class of rank-2 tensors: symmetric (or Hermitian) positive-definite forms.

As with linear applications, represented by matrices, tensors may be applied to elements of vector spaces, giving rise the following:

Definition 3 (Application). *The application of a rank-k tensor ϕ, represented by the matrix $M \in \mathbb{M}_{i_1 i_2 \ldots i_k}(\mathbb{F})$, to a vector $v \in \mathbb{V}$ is a rank $k-1$ tensor represented by the matrix $\widetilde{M}$:*

$$\widetilde{M}_{i_1 \ldots i_{j-1} i_{j+1} \ldots i_k} := M_{i_1 \ldots i_j \ldots i_k} v_{i_j} . \tag{3}$$

The reader should notice that the previous definition of *application* is a generalization of the common matrix-vector product. In other terms, applying a tensor to a vector simply "eats" one dimension, as evidenced by the indices in the previous definition.

Following a standard algebraic practice, indices may be rendered more intelligible when dealing with a tiny number of dimensions, *e.g.*, do not exceed 4 dimensions: in this case, instead of using the more correct, yet more complex, form of $M_{i_1 i_2 i_3 i_4}$, we will employ the clearer notation of M_{ijkl}.

Example 3. Let us consider a rank-3 tensor ϕ represented by the matrix $M \in \mathbb{M}_{333}(\mathbb{R})$ and a vector $v \in \mathbb{R}^3$:

$$M = \left(\begin{pmatrix} 1 \\ 1 \\ 1 \\ 2 \\ 2 \\ 2 \\ 3 \\ 3 \\ 3 \end{pmatrix} \begin{pmatrix} 1 \\ 1 \\ 1 \\ 2 \\ 2 \\ 2 \\ 3 \\ 3 \\ 3 \end{pmatrix} \begin{pmatrix} 1 \\ 1 \\ 1 \\ 2 \\ 2 \\ 2 \\ 3 \\ 3 \\ 3 \end{pmatrix} \right) , \quad v = \begin{pmatrix} 0 \\ 1 \\ 0 \end{pmatrix} .$$

The result of the application of ϕ to v is a rank-2 tensor

$$\langle \phi, v \rangle = M_{ijk} v_i = \begin{pmatrix} 2\ 2\ 2 \\ 2\ 2\ 2 \\ 2\ 2\ 2 \end{pmatrix} .$$

We mention in passing that applications of tensors span beyond simple vectors, with roots in Grassmann algebra, *i.e.*, the algebra of *multidimensional vectors* (cf. [16]).

Definition 4 (Kroneker Tensor). *The rank-k tensor $\delta_{i_1 i_2 \ldots i_k}$, whose matricial representation is $M_{i_1 i_2 \ldots i_k} = 1$ iff $i_1 = i_2 = \ldots = i_k$, is called Kroneker tensor.*

Kroneker tensor is well known in diverse fields ranging from computer science to economics, and it is commonly known as *Kroneker delta*. Moreover, with the canonical duality between forms and vectors [13], it is common to employ the very same symbol for a vector specification, *e.g.*, $v \in \mathbb{R}^3$, with $v = \delta_2 = (0\ 1\ 0)^t$. Note as the notation δ_2 does not contain any reference to the dimension, due to any lack of ambiguity, since here $\delta_2 \in \mathbb{R}^3$; this notation means that each component in a position different from 2 has value 0, while the component in position 2 has value 1.

The last definition we are going to introduce is a particular restriction of the general concept of maps between spaces, and a common notation in functional programming.

Definition 5 (Map). *The* map *is a function with domain a pair constituted by a function and a vector space, and codomain a second vector space:*

$$\mathrm{map} : \mathcal{F} \times \mathbb{V} \to \mathbb{W}, \tag{4}$$

which associates to each component of a vector, the result of the application of a given function to the component itself:

$$\mathrm{map}(f, v) = w, \quad w_i := f(v_i), \quad f \in \mathcal{F},$$

with $f : \mathbb{V} \to \mathbb{W}$, $v \in \mathbb{V}$, *and* $w \in \mathbb{W}$. *Clearly, the space* $\mathcal{F}$ *is the space of functions with domain* $\mathbb{V}$ *and codomain* $\mathbb{W}$, *i.e.,* $\mathcal{F} = \{\kappa : \mathbb{V} \to \mathbb{W}\}$, *as the reader noted.*

The *map* definition will be exemplified by the following:

Example 4. Let us consider a vectors $v \in \mathbb{R}^3$, with $v = (0 \ ^1/_2 \pi \ \pi)^t$. The *map* of the sin function on v results in

$$\mathrm{map}(\sin, v) = (\sin(0) \ \sin(^1/_2\pi) \ \sin(\pi))^t = \left(0 \ 1 \ 0 \right)^t .$$

4 RFID Data Modeling

This section is devoted to the definition of a general model capable of representing all aspects of a given supply chain.

Our overall objective is to give a rigorous definition of a supply chain, along with few significant properties, and show how such representation is mapped within a standard tensorial framework.

4.1 A General Model

Let us define the set $\mathcal{E}$ as the set of all EPCs, with $\mathcal{E}$ being finite. A *property* of an EPC is defined as an application $\pi : \mathcal{E} \to \Pi$, where Π represents a suitable property codomain.

Therefore, we define the application of a property $\pi(e) := \langle \pi, e \rangle$, i.e., a property related to an EPC $e \in \mathcal{E}$ is defined by means of the pairing EPC-property; a property is a surjective mapping between an EPC and its corresponding property value.

A *supply chain* is defined as the product set of all EPCs, and all the associated properties. Formally, let us introduce the family of properties π_i, $i = 1, \ldots, k + d < \infty$, and their corresponding sets Π_i; we may therefore model a supply chain as the product set

$$\mathcal{S} = \mathcal{E} \times \Pi_1 \times \ldots \times \Pi_{k-1} \times \Pi_k \times \ldots \times \Pi_{k+d} . \tag{5}$$

We highlight the indices employed in the definition of the supply chain. The reader should notice as we divided explicitly the first k spaces $\mathcal{E}, \Pi_1, \ldots, \Pi_{k-1}$, from the remaining ones. As a matter of fact, properties may be split into two different categories: *countable* and *uncountable* ones. Such distinction will be described, and hence utilized, in Section 4.3.

By definition, a set A is *countable* if there exists a function $f : A \to \mathbb{N}$, with f being injective; for example, every subset of natural numbers $U \subseteq \mathbb{N}$ is countable (possibly infinite), and relative and rational sets, $\mathbb{Z}$ and $\mathbb{Q}$, respectively, are countable. The real numbers set $\mathbb{R}$, for instance, is a renowned uncountable set.

4.2 Properties

In the following we will focus on some of the properties related to EPCs, *i.e.*, we will model some codomains Π and their associated features.

Location. Let us briefly model the *location* associated to an EPC. It is common to employ a GPS system in order to track the position on earth, however, any fine-grained space is sufficient to our purposes. In particular, being the earth homeomorphic to a 3-sphere, any ordered triple of real numbers suffices, leading us to the following definition:

Definition 6 (Location). *Let $\mathcal{L}$ be the set of ordered tuples $\ell := (\ell_1, \ell_2, \ell_3)$, with $\ell_1, \ell_2, \ell_3 \in \mathbb{R}$. We name $\mathcal{L}$ as* location set, *ℓ_1, ℓ_2 and ℓ_3 as* location coordinates.

A commonly employed coordinate system is the GPS location, *i.e.*, latitude and longitude, with an additional altitude coordinate. However, in many cases the altitude has no influence from an applicative point of view, thus leading the location to be defined by a couple of real numbers, or better, $\ell = (\ell_{lat}, \ell_{long}, \ell_{alt} \equiv 0)$.

Some standard representation of locations abstract their geographical information, preferring to employ a simple mapping between locations and a subset of natural numbers, representing the *location identifier*. Such expression is in contrast with the previous definition with respect to the *countability* property: employing a coordinate system will make location an uncountable property, while electing an identifier lets the property set fall in the countable category.

Which representation is more suitable to an application, is a matter of choice with respect to the domain of the problem, and therefore beyond the scope of this manuscript.

Time. In order to model a temporal interval relative to a product (EPC), we resort to an ordered couple of elements from the ring of real numbers. This suffices to specify, *e.g.*, the *entry* and *exit* time of a product from a given location. With this point of view, we model time as follows:

Definition 7 (Time). *Let $\mathcal{T}$ be the set of ordered couples $\tau := (t_i, t_o)$, with $t_i, t_o \in \mathbb{R}$. We name $\mathcal{T}$ as* time set, *t_i and t_o as* incoming *and* outcoming *timestamps,* respectively.

We highlight the fact that time spaces need not to be modeled as real numbers: in fact, subsets of $\mathbb{R}$ such as natural numbers, may also be suitable within a particular context, *e.g.*, employing UNIX timestamps. However, our model aims at generality, and therefore the real ring is the most appropriate choice, being the natural numbers set a proper subset of the real numbers ring.

Definition 8 (Inner sum). *Let us define the* inner sum *of two time elements* $\tau_1 = (t_i^1, t_o^1)$, $\tau_2 = (t_i^2, t_o^2)$ *as the operator* $\oplus : \mathcal{T} \times \mathcal{T} \to \mathcal{T}$:

$$\tau_1 \oplus \tau_2 := \left(\min(t_i^1, t_i^2), \max(t_o^1, t_o^2) \right) .$$

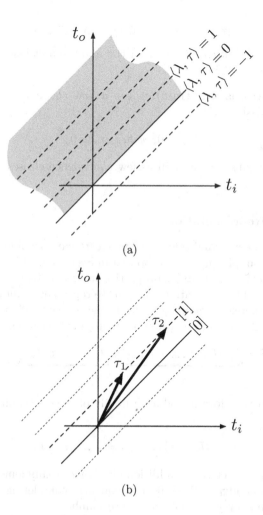

(a)

(b)

Fig. 1. The admissible time $\langle \lambda, \tau \rangle > 0$ region pictured in red is shown in (a); in (b) the two times τ_1 and τ_2 belong to the same equivalence class, *i.e.*, [1]

Such operation allows us to rigorously model the "addition of products", as in a product assembly line: a paint bucket, for instance, is the addition of the metallic container and liquid dye, hence the above definition of time summation. With the above operator, we have that $(\mathcal{T}, \oplus)$ assumes the algebraic structure of an abelian group.

Definition 9 (Lifetime). *We define as* lifetime *the linear form* $\lambda : \mathcal{T} \to \mathbb{R}$ *defined as follows:*

$$\langle \lambda, \tau \rangle := t_o - t_i \,, \quad \tau = (t_i, t_o) \in \mathcal{T} \,.$$

Due to the linearity, we are allowed to construct equivalence classes in $\mathcal{T}$ as follows:

$$[\widetilde{\tau}] := \{ \tau \in \mathcal{T} : \langle \lambda, \tau \rangle = \langle \lambda, \widetilde{\tau} \rangle \} \,. \tag{6}$$

The *canonical representative elements* of the above equivalence classes are defined as $(0, t_o)$, with $t_o \in \mathbb{R}$.

Definition 10 (Admissible time). *Given a time element* $\tau \in \mathcal{T}$*, we say that* $\tau = (t_i, t_o)$ *is* admissible *iff*

$$\langle \lambda, \tau \rangle > 0, \; with \; t_i, t_o > 0 \,.$$

Both admissibility subspace, as well as few equivalence classes, are pictured in Fig. 1.

4.3   Tensorial Representation

Let us now introduce a formal *tensorial framework* capable of grasping all properties related to a supply chain, as proposed in Section 4.1.

As previously outlined, we divide properties into two categories, *countable* and *uncountable* spaces. This separation allows us to represent countable spaces with natural numbers, therefore mapping their product space to $\mathbb{N}^k$, while leaving the product space of all uncountable properties into a collective space $\mathbb{U}$:

$$\mathcal{S} = \underbrace{\mathcal{E} \times \Pi_1 \times \ldots \times \Pi_{k-1}}_{\mathbb{N}^k} \times \underbrace{\Pi_k \times \ldots \times \Pi_{k+d}}_{\mathbb{U}} \,. \tag{7}$$

Such mapping will therefore introduce a family of injective functions called *indexes*, defined as:

$$\mathrm{idx}_i : \Pi_i \longrightarrow \mathbb{N}, \qquad i = 1, \ldots, k - 1 \,. \tag{8}$$

When considering the set $\mathcal{E}$, we additionally define a supplemental index, the *EPC index function* $\mathrm{idx}_0 : \mathcal{E} \to \mathbb{N}$, consequently completing the map of all countable sets of a supply chain $\mathcal{S}$ to natural numbers.

It should be hence straightforward to recognize the tensorial representation of a supply chain:

Definition 11 (Tensorial Representation). *The* tensorial representation *of a supply chain* $\mathcal{S}$, *as introduced in equation* (5), *with countability mapping as in* (7) *is a multilinear form*

$$\Sigma : \mathbb{N}^k \longrightarrow \mathbb{U}. \tag{9}$$

A supply chain can be therefore rigorously denoted as a rank-k tensor with values in $\mathbb{U}$, mapping countable to uncountable product space.

4.4 Implementation

Preliminary to describing an implementation of our supply chain model, based on tensorial algebra, we pose our attention on the practical nature of a supply chain.

Our treatment is general, representing a supply chain with a tensor, *i.e.*, with a multidimensional matrix. Several other approaches are available in literature on the same topic, with a graph being the preferred supply chain model [17]. These models are thoroughly represented by our tensorial approach, due to the well known mapping between graphs and matrices (cf. [18], [19], and §1.2 of [20]).

However, a matrix-based representation need not to be *complete*. Supply chain rarely exhibit completion: practical evidence [17] suggests that products, identified by their EPC, for example, seldom present themselves in every location. The same consideration applies also to other properties, in particular, to countable properties.

Hence, our matrix effectively requires to store only the information regarding connected nodes in the graph: as a consequence, we are considering sparse matrices [21], *i.e.*, matrices storing only non-zero elements.

Notation. A sparse matrix may be indicated with different notations (cf. [22] and [21]), however, for simplicity's sake, we adopt the *tuple notation*. This particular notation declares the components of a tensor $M_{i_1 i_2 \ldots i_k}(\mathbb{U})$ in the form of

$$\mathcal{M} = \left\{ \{i_1 i_2 \ldots i_k\} \to u \neq 0\, , u \in \mathbb{U} \right\}, \tag{10}$$

where we implicitly intended $u \neq 0 \equiv 0_{\mathbb{U}}$. As a clarifying example, consider a vector $\delta_4 \in \mathbb{R}^5$: its sparse representation will be therefore constituted by a single tuple of one component with value 1, *i.e.*, $\{ \{4\} \to 1 \}$.

Example 5. Let us consider the supply chain pictured in Fig. 2, described tensorially by $\Sigma : \mathbb{N}^2 \to \mathcal{T}$, whose representative matrix is as follows:

$$\begin{pmatrix} (0,2) & \cdot & \cdot & (5,7) & \cdot & (8,10) & \cdot & \cdot \\ \cdot & \cdot & (0,3) & \cdot & \cdot & \cdot & \cdot & (4,9) \\ (0,1) & \cdot & \cdot & (4,5) & \cdot & \cdot & (6,9) & \cdot \\ \cdot & (0,4) & \cdot & \cdot & (5,6) & \cdot & \cdot & (8,9) \\ \cdot & \cdot & (0,5) & \cdot & (6,7) & \cdot & \cdot & (9,11) \end{pmatrix}$$

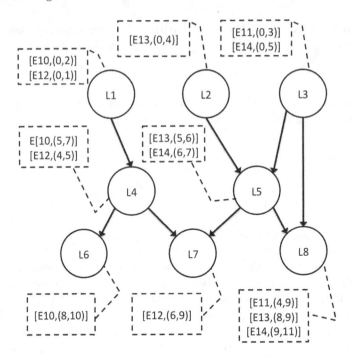

Fig. 2. An example supply chain represented by a directed graph. Callouts represent lists of tuple constituted by an EPC with the associated time (t_i, t_o).

where, for typographical simplicity, we omitted $0_\mathcal{T} = (0, 0)$, denoted with a dot.

For clarity's sake, we outline the fact that Σ is a rank-2 tensor with dimensions 5, 8, *i.e.*, the number of rows and columns, respectively. In fact, the EPC and locations sets are as follows:

$$\mathcal{E} = \{E10, E11, E12, E13, E14\},$$
$$\Pi_1 = \{L1, L2, L3, L4, L5, L6, L7, L8\}.$$

where $\mathrm{idx}_0(E10) = 1$, $\mathrm{idx}_0(E11) = 2$, ..., $\mathrm{idx}_0(E14) = 5$, and similarly $\mathrm{idx}_1(L1) = 1, \ldots, \mathrm{idx}_1(L8) = 8$. In the sparse representation we have $\{\{1, 1\} \rightarrow (0, 2), \{1, 4\} \rightarrow (5, 7), \ldots, \{5, 8\} \rightarrow (9, 11)\}$.

Data Structure. Several data structures have been proposed for (sparse) tensor representations, which in computer science are usually known as *multidimensional arrays* or *n-arrays* [23].

A common representation for sparse matrices is the *Compressed-Row Storage* format, or briefly CRS, with its dual *Compressed-Column Storage* (CCS) counterpart (cf. [21], [20], and [22]). Such matrices with nnz non-zero entries, are represented by means of three different arrays: one of length nnz representing all stored entries, in row-major order for CRS (column-major for CCS); one array of length equal to the number of rows, containing the indexes of the first element of each row (or column for CCS); finally, the column vector of length nnz, storing the column index (or row for CCS) of each non-zero element.

Literature describes numerous data structures related to CRS, aimed at tensor representation. In essence, elements are stored by *sorting* indexes first, and subsequently memorizing index vectors as described above, *e.g.*, for a rank-3 tensor T_{ijk}, storing CRS matrices corresponding to each constant dimension, for example T_{1jk}, T_{2jk}, and T_{3jk}, a technique commonly known as *slicing*. However, all CRS descendants suffer from the same drawbacks of their ancestor: they are tailored for specific tensor ranks, specifically up to 4th rank multilinear forms for specialized data structures, and therefore cannot be employed when ranks may vary, or with a discrete number of dimensions. A thorough discussion of sparse tensor data structures and their technical aspects is beyond the scope of this manuscript, and for more details we refer the reader to [24], [25], and [26].

Our objective is to represent tensors of any rank, with the highest possible flexibility: we therefore chose another common data format, the *Coordinate Sparse Tensors* [27], or CST.

This data structure, as pictured in the ensuing paragraph, consists of a list of nnz entries, describing the entry value and coordinates, parallel to the description given in Section 4.4. The main advantage of this organization is simplicity and adaptability, discarding the previous constraint on dimensional sorting. As we will prove in Section 7, this simple format exhibits almost linear complexity for the operations described in Section 3, and utilized extensively in Sections 5 and 6.

4.5   Remarks

The reader noticed that all spaces, and their correlated operations, as defined in Sections 3 and 4, exhibit fundamental algebraic properties: *associativity*, *commutativity*, and *distributivity*. These characteristics help us in clarifying implementative aspects regarding our sparse approach: memory constraints, and parallel computations.

At first sight, in order to perform operations such as tensor application, we may need to load the whole matrix and afterwards execute the required computation. Due to the distributivity of products with respect to summations, we actually may *split* a matrix in sub-matrices, carry out the application and finally, thanks to the associativity, add the partial results obtaining the correct solution.

For example, if $A \in M_{22}(\mathbb{R})$, and $v \in \mathbb{R}^2$ being a suitable vector, the matrix can be divided in two (or more) sub-matrices, thus giving the ability to perform computations independently, as

$$Av = (A' + A'')v = A'v + A''v \tag{11}$$

The above equation shows how the *sparsity* of a matrix—*i.e.*, the ratio between the overall dimensions and the actual non-zero stored elements—alleviates problems caused by the finiteness of computer memories. When performing $A'v$, we actually do not need any information regarding the subsequent A'' matrix and vice versa, and as a result, we may not store its elements in memory.

On one hand, we are capable of making use of our framework in memory-constrained systems; on the other hand, the very same consideration may be applied in carrying the above computation on parallel systems.

Owing to the distributive property, as previously outlined, the list of non-zero items that constitute the representation of a given tensor may be split among available processors. As an example, let $T \in M_{i_1 i_2 \ldots i_k}$ be a rank-k tensor, and let nnz be the number of stored elements of T. It is straightforward to subdivide *element-wise* [28] the items as follows:

$$
\left.
\begin{array}{lll}
v_1 & t_{i_1}^{1} & \ldots \ t_{i_k}^{1} \\
v_2 & t_{i_1}^{2} & \ldots \ t_{i_k}^{2} \\
\vdots & & \\
v_{nnz/p} & t_{i_1}^{nnz/p} & \ldots \ t_{i_k}^{nnz/p}
\end{array}
\right\} p_1
$$

$$
\left.
\begin{array}{lll}
v_{nnz/p+1} & t_{i_1}^{nnz/p+1} & \ldots \ t_{i_k}^{nnz/p+1} \\
v_{nnz/p+2} & t_{i_1}^{nnz/p+2} & \ldots \ t_{i_k}^{nnz/p+2} \\
\vdots & & \\
v_{2nnz/p} & t_{i_1}^{2nnz/p} & \ldots \ t_{i_k}^{2nnz/p}
\end{array}
\right\} p_2
$$

$$
\vdots
$$

$$
\left.
\begin{array}{lll}
v_{(p-1)nnz/p+1} & t_{i_1}^{(p-1)nnz/p+1} & \ldots \ t_{i_k}^{(p-1)nnz/p+1} \\
v_{(p-1)nnz/p+2} & t_{i_1}^{(p-1)nnz/p+2} & \ldots \ t_{i_k}^{(p-1)nnz/p+2} \\
\vdots & & \\
v_{nnz} & t_{i_1}^{nnz} & \ldots \ t_{i_k}^{nnz}
\end{array}
\right\} p_p
$$

with p the number of processors, $p_1, \ldots, p_p$ representing the available processors, $v_u \in \mathbb{U}$ the tensor values, and $t_{i_1}^{q} \ldots, t_{i_k}^{q}$ the *coordinates* referring to the q-th element.

We mention in passing that a *row-wise* subdivision—or in case of rank greater than 2, *slicing* the tensor—has a nefarious effect on the overall performance: in fact, as the reader noticed, rows may not contain the same number of non-zero elements, and therefore some processors, when computing an operation by rows, may undergo a computational stress, while others could remain idle. For an extensive treatise on parallel tensor operations, we suggest the reader [29] and [30].

Literature provides more complex data structures than CRS, CCS, or CST. However, these are mainly tailored for rank-2 and rank-3 tensors [25, 31], and therefore are unsuitable to our purpose: we aim at generality and flexibility, and hence restricting the scope of our supply chain space $\mathcal{S}$ is undesirable.

5   RFID Massive Data Analysis

The ensuing section is devoted to the analysis of massive amount of RFID data via our novel conceptual framrework.

For simplicity's sake, while maintaining generality, we simplify our notation, employing two countable sets, EPC $\mathcal{E}$ and location $\mathcal{L}$ indexes, in that order, and the uncountable time set $\mathcal{T}$, *i.e.*, a matrix M_{ij}: i referring to EPCs, and j being the index of locations. Our examples, in order to illustrate the queries, will be based on the ones provided in Section 4.

Such choice of attributes introduce no limitations on the generality of our approach, due to the fact that, in essence, applying a tensor to a vector means summing all the components sharing an index, regardless of ranks and dimensions.

5.1   Tracking Query

A *tracking query* finds the movement history for a given tag identifier $e \in \mathcal{E}$. We can comfortably perform the query efficiently, using the model described in Section 4, by applying the tensor application.

Therefore, given $i = \text{idx}(e)$, we build a Kroneker vector δ_i, with $|\delta_i| = |\mathcal{E}|$, and finally apply of the rank-2 tensor represented by M_{ij} to δ_i, *i.e.*:

$$r = M_{ij}\delta_i .$$

For instance, referring to the example pictured in Fig. 2, let us consider the tag $E13$, we have $i = \text{idx}(E13) = 4$, and therefore our vector will be $\delta_4 = \{\{4\} \to 1\}$. Consequently, the resulting vector will be

$$r = M_{ij}\delta_4 = \{\{2\} \to (0,4), \{5\} \to (5,6), \{8\} \to (8,9)\}$$

or in another notation, $L2 \to L5 \to L8$.

5.2   Path Oriented Query

A *path oriented query* returns the set of tag identifiers that satisfy different conditions. Following the query templates given in [7], we subdivide path oriented queries into two main categories: *path oriented retrieval* and *path oriented aggregate* queries.

The former returns all tags covering a path satisfying given conditions, while the latter computes an aggregate value. It is possible to formulate a grammar for these queries, similar to XPath expressions; in particular, path oriented retrieval and path oriented aggregate queries are respectively indicated as follows:

$$L_1[\text{cond}_1]//L_2[\text{cond}_2]//\ldots//L_n[\text{cond}_n] , \tag{12a}$$

$$L_1[\text{cond}_1]/L_2[\text{cond}_2]/\ldots/L_n[\text{cond}_n] , \tag{12b}$$

where $L_1, \ldots L_n$ is the *path condition*, *i.e.*, the sequence of locations covered by the tag, and cond_i is the *info condition*; for more information about such expressions, we refer the reader to [7]. A path condition expresses *parenthood* between locations, indicated as L_i/L_j, or *ancestry* with $L_i//L_j$; an info condition, on the other hand, indicates conditions on tag properties, *e.g.*, *StartTime* and *EndTime*.

Path Oriented Retrieval. In our framework, a path oriented retrieval query as in (12a) is easily performed by exploiting the tensor application coupled with the Hadamard product. Given the location set $\mathcal{L}$, for each $z = \mathrm{idx}(L_k)$, with $k = 1, \ldots, n$, we create a Kroneker vector δ_z and subsequently apply the rank-2 tensor, resulting in a set of vectors $r_z = M_{ij}\delta_j$, where for typographical reasons, we dropped the subscript intending $\delta \equiv \delta_z$.

Finally, we apply the condition function to each r_z employing the map operator. This yields a set of vectors

$$\widetilde{r_z} = \mathrm{map}(\mathrm{cond}_z, r_z), \qquad \mathrm{cond}_z : \mathbb{N}^k \times \mathbb{U} \longrightarrow \mathbb{F},$$

whose Hadamard multiplication generates the final result:

$$\bar{r} = \widetilde{r_1} \circ \ldots \circ \widetilde{r_n},$$

reminding the reader that only non-zero values are stored, and therefore given as a result of a computation. The cond functions, as indicated above, are maps between properties of supply chains and a suitable space $\mathbb{F}$, *e.g.*, natural numbers for a boolean result.

For an example, referring to Fig. 2, let us consider the query

$$L3[StartTime > 0]//L8[EndTime - StartTime < 4].$$

In this case, given $\mathrm{idx}(L3) = 3$ and $\mathrm{idx}(L8) = 8$, we build δ_3 and δ_8, and generate the partial results

$$r_3 = M_{ij}\delta_3 = \{\{2\} \to (0,3), \{5\} \to (0,5)\}$$

$$r_8 = M_{ij}\delta_8 = \{\{2\} \to (4,9), \{4\} \to (8,9), \{5\} \to (9,11)\}$$

where evidently the application was performed along the second dimension, *i.e.*, for each $\delta \in \{\delta_3, \delta_8\}$, we compute $M_{ij}\delta_j$. Finally, subsequent to mapping conditions on the results, in this case $\mathrm{cond}_3 = t_i(\cdot) > 0$ and $\mathrm{cond}_8 = \langle \lambda, \cdot \rangle < 4$, we obtain the correct outcome $\bar{r} = \widetilde{r_3} \circ \widetilde{r_8} = \{\{5\} \to (9,11)\}$, *i.e.*, $E14$.

Considering parenthood as in (12b) instead of ancestry, *i.e.*, L_i/L_j, we briefly sketch the fact that such query does not, in fact, differ from the above, since parenthood is a *restriction* on the ancestry relation: each resulting EPC, when subject to a tracking query, produces a sparse vector whose *length*, *i.e.*, the number of non-zero stored elements, is exactly equal to the number of locations under analysis.

Path Oriented Aggregate. Path oriented aggregate queries may be represented as $\langle f, Q \rangle =: f(Q)$ where f is an *aggregate function*, *e.g.*, average or minimum, and Q is the result of a path oriented retrieval query.

Therefore, let $r_{\bar{Q}}$ be the result of Q, and let f be a function defined on vectors of supply chain elements, we simply have that a path aggregate may be expressed as $f(r_{\bar{Q}})$.

Referring to Fig. 2, let us consider the query expressed in the grammar of [7]

$$\langle AVG[L8.StartTime], //L8 \rangle .$$

The above is easily translated into our tensorial framework by applying the function $f := \text{average}(t_i)$ to the outcomes of the path oriented retrieval query on $L8$, resulting in $\bar{r} = \text{average}(\{4, 8, 9\}) = 7$.

6   RFID Decentralized Data Processing

This section is devoted to the exemplification of our conceptual method, with the applicative objective of analyzing RFID data in a decentralized processing. For grater coherence, we use the same simplified notation as in the previous section.

6.1   Monitoring Queries

Usually real-time analysis on a supply chain should monitor the correct traversing of products. For instance, we would control if RFID readers are producing incorrect data, e.g., $t_i > t_o$, that is if an object leaves a location before entering in it, or if the transporting of products has an unexpected delay, or any additional property exceeds a chosen threshold value. We call such analysis a *monitoring query*.

In this case we have to perform quick decentralized on-line processing, *i.e.*, different data sources send local elaborations to be integrated. As we discussed in Section 4, and subsequently proved in Section 8, our framework is able to quickly import RFID data, perform monitoring query efficiently, and finally, thanks to the associativity of the operations, add the partial results obtaining the correct solution.

Lifetime Admissibility. First of all, we introduce the *lifetime admissibility test*, selecting all EPCs whose time is not admissible (cf. Definition 10 in Section 4):

$$\text{map}(\langle \lambda, \cdot \rangle < 0, M_{ij}) . \tag{13}$$

The reader should notice a shorthand notation for an implicit boolean function: as for all descendant of the C programming language, such definition yields 1 if the condition is met, *i.e.*, the time associated to an element of the tensor is not valid, 0 otherwise.

Furthermore, we recall the fact that being M_{ij} represented as a sparse matrix, the application of our condition is well defined, being applicable to the stored values, *i.e.*, to the time couple $(t_i, t_o) \in \mathcal{T}$. For instance, let us suppose the triples $\text{tuple}_1 = (E15, L3, 3)$ and $\text{tuple}_2 = (E15, L3, 2)$ where tuple_1 is generated before than tuple_2. The compression will result a stay record $(E15, L3, 3, 2)$ that does not satisfy the admissibility test, and the error in admissibility will be detected instantly.

Delay. The *delay test* chooses all EPCs whose time surpasses an expected interval. In our framework, given two times τ_1 and τ_2, *i.e.*, the bounds of the interval, we would return all EPCs whose time τ excesses the inner sum $\tau_3 = \tau_1 \oplus \tau_2$ (cf. Definition 8 in Section 4):

$$\text{map}(\, (t_i(\cdot) > t_i(\tau_3)) \wedge (t_o(\cdot) < t_o(\tau_3)) \,, M_{ij}) \,. \tag{14}$$

As an example, let us consider $\tau_1 = (7,8)$ and $\tau_2 = (9,10)$, and hence $\tau_3 = \tau_1 \oplus \tau_2 = (7,10)$: in this case the delay test will return the EPC $E13$. In the same way, it is straightforward to extend similar analysis on other properties of a product.

Last Location. Additionally, in a decentralized environment, we can process retrieved data querying for the path a given EPC is following. In particular, we may introduce the *last location* query, which shows, for a given EPC with index value i, the last known location. This is comfortably formalized in our framework as $\max_{t_i}(\, M_{ij}\delta_i\,)$. In other words we select the location with the maximum value of t_i for the given EPC. In our example, the last known location for EPC $E13$ would be $L8$.

Common Location. Additionally, we may supervise some properties regarding diverse locations or EPCs. We mention in passing a *common test*, that given two locations $L1$ and $L2$, finds all the common EPCs: given $j = \text{idx}(L1)$ and $k = \text{idx}(L2)$, the result is given by $M_{ij}\delta_j \circ M_{ik}\delta_k$. As an example, let us consider $L4$ and $L6$ as in Fig. 2, with indexes 4 and 6, respectively. The common EPCs are therefore given by $M_{ij}\delta_4 \circ M_{ij}\delta_6$, hence

$$\{\{1\} \to (5,7), \{3\} \to (4,5)\} \circ \{\{1\} \to (8,10)\}$$

which yields the correct result, selecting only the EPC $E10$ with index 1. The reader should notice that in the above examples, and in the following ones, the subscript for the index function has been dropped, due to its unambiguity, *e.g.*, $\text{idx}(E10) = \text{idx}_0(E10)$.

6.2   Error Detection Queries

Detecting errors is a major objective of RFID data monitoring. We may easily spot in the previous paragraph a simple error detection query in the admissibility test. Such examination, in fact, shows all readings that yield an incorrect lifespan, *i.e.*, a reading that shows that an EPC has moved outside a location before entering in it.

General Admissibility. It is straightforward to generalize the lifetime admissibility test of Equation (13) for any given EPC property. Let us substitute the lifetime pairing $\langle \lambda, \cdot \rangle$ with a suitable given function

$$\text{threshold} : \mathbb{U} \times \mathbb{R}^k \longrightarrow \mathbb{R} \,,$$

where $\mathbb{R}^k$ allows the definition of complex spaces and conditions, *e.g.*, of intervals with $\mathbb{R}^2$. As the reader will notice, admissibility tests can be performed exclusively on uncountable properties.

Path Consistency. Path-focused errors are more arduous to be detected. A *path consistency* query recognizes, for a given EPC, if the followed path is consistent with timings, *i.e.*, if it accesses a location prior to departing the previous one.

Let us introduce the path-consistency function $\mathcal{P}$ defined as

$$\mathcal{P} : \mathcal{T} \times \mathcal{T} \longrightarrow \mathbb{R}, \tag{15}$$

that yields an invalid time value if two successive times $\tau_1, \tau_2 \in \mathcal{T}$ are not consistent as previously stated. Formalizing the definition, we have

$$\mathcal{P}((t_i, t_o)^1, (t_i, t_o)^2) = \begin{cases} (t_i, t_o)^1, & t_o^1 \leq t_i^2, \\ (-\infty, +\infty), & \text{otherwise}. \end{cases} \tag{16}$$

For example, let us show a valid time comparison, *e.g.*, $\mathcal{P}((1,5),(6,10)) = (1,5)$. The purpose of such definition will be clear in the following paragraph.

Hence, given $i = \text{idx}(e)$, being $e \in \mathcal{E}$ the elected EPC, in order to control whether e followed a consistent path, let us find all the relative locations *sorted* by their incoming timestamp t_i:

$$\text{sort}_{t_i}(M_{ij}\delta_i). \tag{17}$$

Therefore, by successive application of the $\mathcal{P}$ function on the preceding result, itself being a sparse matrix, we attain the objective of detecting path inconsistency. We underline the fact that $\mathcal{P}$ is an *associative* function, without the commutativity property. The result will consequently be obtained by *folding* the function $\mathcal{P}$ on the sorted sparse partial solution followed by the time $\tau_\infty = (+\infty, -\infty)$:

$$\text{fold}\left(\mathcal{P}, \text{sort}_{t_i}(M_{ij}\delta_i) \cup \{(+\infty, -\infty)\}\right). \tag{18}$$

The *fold* or *accumulation* is a higher-order function that, given a function and a (non empty) list, recursively iterates over the list elements, applying the input function to each element and the preceding partial result; a comprehensive formalization of folding is beyond the scope of this manuscript, being such function a cornerstone of functional programming; for more information we refer the reader to [32] and [33].

Consequently, given Equation (18), the folding proceeds as follows:

$$\mathcal{P}((t_i, t_o)_1, \mathcal{P}((t_i, t_o)_2, \ldots, \mathcal{P}((t_i, t_o)_n, (+\infty, -\infty))\ldots))).$$

The above computation is advantageously visualized as a *computation binary tree*, where leaves contain a list element, and intermediate nodes indicate the folded function:

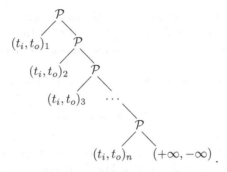

In order to illustrate the *path consistency* query, we should focus on EPC $E13$ from the example pictured in Fig. 2. The result of Equation 17, given also in the preceding subsection *Monitoring Queries*, is

$$S := \{\{2\} \rightarrow (0, 4), \{5\} \rightarrow (5, 6), \{8\} \rightarrow (8, 9)\} .$$

Finally, we need to apply recursively $\mathcal{P}$ to the list (sparse matrix) via *folding*:

$$
\begin{aligned}
\mathrm{fold}(\mathcal{P}, S) &= \mathcal{P}((0, 4), \mathcal{P}((5, 6), \mathcal{P}((8, 9), (+\infty, -\infty)))) \\
&= \mathcal{P}((0, 4), \mathcal{P}((5, 6), (8, 9))) \\
&= \mathcal{P}((0, 4), (5, 6)) \\
&= (0, 4) .
\end{aligned}
$$

Let us consider now an invalid path $\widetilde{S}$, where $E13$ enters location 8 with timestamp 5, *i.e.*, $\{8\} \rightarrow (5, 9)$. The computation above will therefore yield an invalid value, as expected:

$$
\begin{aligned}
\mathrm{fold}(\mathcal{P}, \widetilde{S}) &= \mathcal{P}((0, 4), \mathcal{P}((5, 6), \mathcal{P}((5, 9), (+\infty, -\infty)))) \\
&= \mathcal{P}((0, 4), \mathcal{P}((5, 6), (5, 9))) \\
&= \mathcal{P}((0, 4), (-\infty, +\infty)) \\
&= (-\infty, +\infty) .
\end{aligned}
$$

6.3 Cleaning Queries

In our framework cleaning queries can be considered as *insertion* and *deletion* operations on the matrix. In particular our framework is based on a tensorial representation of RFID data, based on stay records. This allows us not only to express EPCs and their properties within a theoretically sound mathematical model, but also to implicitly discarding duplicate records from diverse RFID readers. Moreover, by making use of tensor representations, we exploit the underlying computational capabilities of modern processors, natively designed for integer and floating point operations.

In order to eliminate duplicates, and removing erroneous data, *e.g.*, inconsistent paths, insertion and deletion queries are pivotal functions that must be implemented. As a matter of fact, in our computational framework these procedures are reflected by *assigning values* in a given tensor. Removing a value is comfortably implemented by assigning the null value to an element, *i.e.*, $M_{ij} = 0_{\mathcal{T}}$, when referring to our example (cf. Fig. 2), and analogously, inserting or modifying a value is attained via a simple operation $M_{ij} = (t_i, t_o)$. Most specifically when a raw stream, (e, l, t), is generated if e does not exist in l, we insert a new stay record (e, l, t, t)—that is a *basic insertion*—otherwise, we update the t_o value of e in l with t—that is an *update*.

7   Theoretical Analysis

As previously outlined in Section 4.4, our representation of sparse rank-k tensors is based on the Coordinate Sparse Tensors, consisting of tuples constituted by the *entry value* $u \in \mathbb{U}$, and its corresponding *coordinates* $(v_1, \ldots, v_k) \in \mathbb{N}^k$, according to Section 4.3.

The ensuing paragraphs analyze the theoretical complexity of all the operations involved in RFID queries, according to Sections 5 and 6.

In the following, we will employ the notation $nnz(T)$, with T being the rank-k tensor under analysis, denoting the number of non-zero values of the multilinear form. Analogously, when dealing with a sparse vector, $nnz(v)$ will yield the number of non-zero entries of v: the same notation, nnz, will introduce no ambiguity, as arguments of nnz are uniquely identified either with tensors or with vectors, in uppercase and lowercase letters, respectively.

Insertion. The assembly of a sparse tensor requires the basic operation of inserting an element into the list of non-zero values. Clearly, this action only involves controlling if such element is already present, therefore the operation has a complexity of $O(nnz(T))$. The same analysis applies to sparse vectors, with an asymptotic complexity of $O(nnz(v))$.

Deletion and Update. Such basic actions mimic the above insertion operation, and therefore have an asymptotic complexity of $O(nnz(T))$.

Hadamard Product. The Hadamard product of two vectors $u \circ v$ has a complexity of $O(nnz(u)nnz(v))$. This trivial operation, multiplies each value of the first vector u for its corresponding element of the second vector v, if it exists. By simply storing values *ordered* by their coordinates, the complexity may be reduced to $O(nnz(u))$.

Tensor Application. For a suitable vector v, the tensor application on the i_j-th dimention $T_{i_1 i_2 \ldots i_k} v_{i_j}$, begin T a rank-k tensor, has asymptotic complexity of $O(nnz(T))$, as detailed in [27].

Map and Fold. Both *mapping* a function on a vector, or *folding*, have asymptotic complexity of $O(nnz(v))$, visibly depending on the number of non-zero elements; actual computations on zero values are performed only once, and the

Table 1. Parameters for synthetic data generation

Grouping Factor	(50, 20, 10, 5, 3, 1, 1, 1)
Minimum Path Length	4
Maximum Path Length	8
Moving Rate	0.5

outcome is stored as the default value for the resulting vector. It is straightforward to compute the asymptotic complexity of mapping functions on tensors, resulting in $O\left(nnz(T)\right)$.

We stress the fact that the Coordinate Sparse Tensor data structure is extremely flexible, while retaining sufficient simplicity. For a thorough review of operations on sparse tensors and vectors, their asymptotic complexities and theoretical proofs, we refer the reader to [21], [27], and [23].

8 Experiments

We performed a series of experiments aimed at evaluating the performance of our approach (denoted as T), reporting the main results in the present section.

Our benchmarking system is a dual core 2.66 GHz Intel with 2 GB of main memory running on Linux, where we implemented our framework[1] in C++ within the Mathematica 8.0[2] computational environment. Our results have been compared to the ones from the approach in [7], tested against a generated synthetic RFID data in terms of stay records, and considering products moving together in small groups or individually, a behavior called *IData* in [7]: data production followed the same guidelines on a supply chain of 100 locations. In Table 1 we show the parameters used to generate the synthetic IData data.

A *grouping factor* $(g_1, g_2, \ldots, g_k)$ signifies that in the i-th stage, the number of products in a group is g_i. The generation for the next location is stopped, or the next location is generated, according to the *moving rate*. The time information in stay records is randomly generated. Finally the complete data set comprises 10^5, $5 \cdot 10^5$, 10^6, $5 \cdot 10^6$, and 10^7 stay records.

The performance of the systems has been measured with respect to: (i) data loading, (ii) massive analysis, and (iii) decentralized query execution. In particular, for both analysis and query execution, we performed *cold-cache* experiments, *i.e.*, dropping all file-system caches before restarting the systems and running the queries, and repeated all the tests three times, reporting the average execution time.

---

[1] A prototype implementation is available online as Web Service at
http://pamir.dia.uniroma3.it:8080/SimpleWebMathematica/
[2] http://www.wolfram.com/mathematica/

8.1 Data Loading

Referring to data loading, the main advantage of our approach is that we are able to perform loading without any particular relational schema, when compared to P, where a schema coupled with appropriate indexes have to be maintained by the system. In this case, loading execution times are 0.9, 11, and 113 seconds, for sets of 10^5, 10^6, and 10^7 stay records, respectively; on the contrary, P timings were in order of minutes and hours. Another significant advantage of T relies in memory consumption: we need 13, 184, and 1450 MB to import the above mentioned sets; as a side-note, we highlight the fact that the 10^7 set required a division in smaller blocks, $e.g.$, 10^6, due to the limited memory at disposal.

8.2 Massive Analysis

In the following, we will denote our tensorial approach T, while the proposed one in [7] with P.

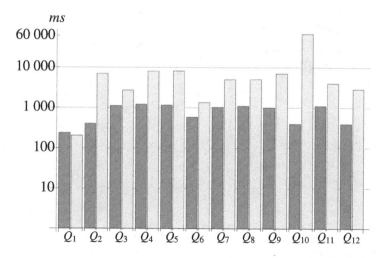

Fig. 3. Query execution time in logarithmic scale for 10^7 stay records: the dark gray bars refer to T, while light gray ones to P

With respect to massive analysis, T presents a similar behavior and advantages with respect to P, for both time and memory consumption. As in [7], we formulated 12 queries to test the two systems as reported in Table 2. In brief, Q1 is a tracking query, Q2 to Q5 are path oriented retrieval queries, while Q6 to Q12 are path oriented aggregate queries. Due to the nature of P, we were able to perform a comparison only on centralized off-line massive analysis.

Fig. 3 shows the query performance times in milliseconds with 10^7 stay records. Both T and P present similar performance to execute tracking queries, $i.e.$, Q1 in the query set: this is normal behavior, since databases present indexes that enhance selection queries, and thus we perform *on par*. With path oriented queries,

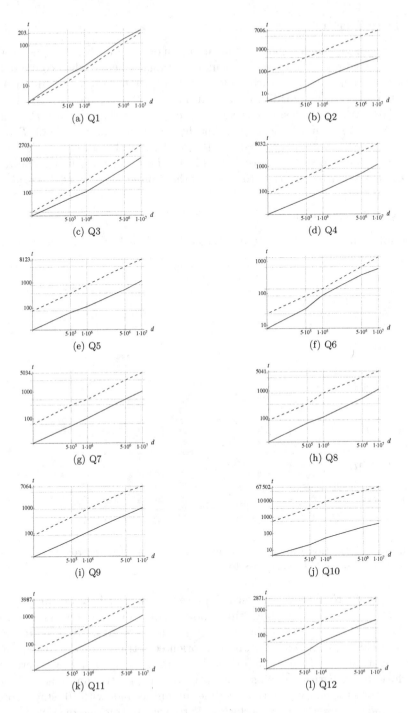

Fig. 4. Query scalability for all data sets as indicated in Section 8 for Q1 (a) to Q12 (l); query execution time (t) in *msec* is plotted, in logarithmic scale, against data set size (d). Dashed graphs refers to *P*, while solid ones to *T*.

Table 2. Test queries expressed using the XPath-like syntax of Lee et. al

Q1	EPC = 17281
Q2	//L628
Q3	/L03/L16/L29/L321/L422
Q4	//L16//L322//L628
Q5	//L16//L322[StartTime<200]//L628
Q6	COUNT(), //L628
Q7	COUNT(), //L16//L322//L628
Q8	COUNT(), //L16//L322[StartTime<200]//L628
Q9	AVG(L16.EndTime - L16.StartTime), //L16//L322//L628
Q10	AVG(L628.StartTime), //L628
Q11	MIN(L16.EndTime - L16.StartTime), //L16//L322//L628
Q12	MIN(L628.StartTime), //L628

as indicated in the plot, T is faster than P: while T executes such queries again as tensor applications, P has to perform SQL queries with different complexities, such as *join operations* for info conditions or aggregations; on average, T executes in the range $[500, 1000]$ $msec$, while P in $[3000, 10000]$ $msec$. In Fig. 4, the reader finds detailed diagrams, showing query performances according to the number of stay records for each of the twelve queries, for all test data sets. For typographical reasons, the first dataset label 10^5 is omitted.

A more significant result is the *speed-up* between the two approaches, as shown in Fig. 5; again, Q1 label is dropped due to typographical causes. We computed the speed-up for all data sets as the ratio between the execution time of P, and that of our approach T, or briefly $S = t_P/t_T$. In general, T performs very well with respect to P in any dataset, particularly for queries related to the object transition, *e.g.*, Q2, Q4, Q5, Q10. The query performance of T is on the average 19 times better than that of P, 150 times on the maximum, *i.e.*, Q10.

Another strong point of T is a very low consumption of memory, due to the *sparse matrix* representation of tensors and vectors. Fig. 6 illustrates the main memory consumption of each query with respect to 10^5, 10^6, and 10^7 stay records. On the average, tracking queries require very few bytes of memory for any dataset, path oriented queries few KBytes, topping 1 MB for 10^7. Results demonstrate how our approach can be used in a wide range of applications, where devices with limited calculus resources may process large amount of data in an efficient and effective way.

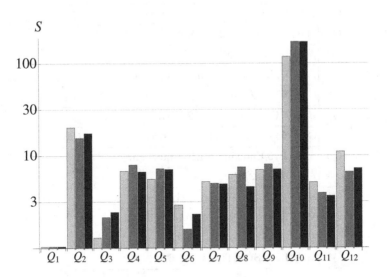

Fig. 5. Speed-up (logarithmic) for all queries with dataset sizes of 10^5 (light gray), 10^6 (dark gray), and 10^7 (black)

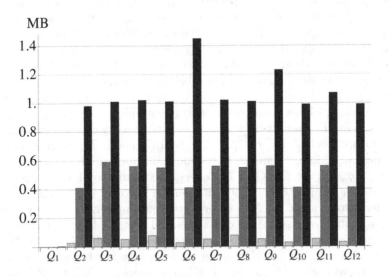

Fig. 6. Main memory consumption for each query in megabytes: light gray bars refer to 10^5, dark gray to 10^6, and black to 10^7 data size

Table 3. Test queries for decentralized query execution

Q1	EPC = 17281
Q2	map($\langle t_o, \cdot \rangle$ < 320, M_{ij})
Q3	$\tau_1 = (0, 5)$ and $\tau_2 = (4, 10)$
Q4	$last(17281)$
Q5	L_1 = L16 and L_2 = L628
Q6	EPC = 17281
Q7	*basic insertion*
Q8	*update*
Q9	*delete*

8.3 Decentralized Query Execution

With respect to massive analysis, T presents a similar behavior and advantages, for both time and memory consumption.

As shown in Table 3, we formulated 9 queries to test the system. In brief, Q1 is a tracking query, Q2 and Q3 are delay and admissibility tests, respectively, Q4 performs the last location, Q5 the common test, Q6 the path consistency, Q7 a basic insertion, Q8 an update and Q9 a deletion (*i.e.*, insertion and update are executed each time a new stream is generated, while deletion is applied to an EPC selected randomly). All these queries were executed as the generation of synthetic data was running. In other words, using the same supply chain of the massive analysis, at periodic times we executed the queries on the data

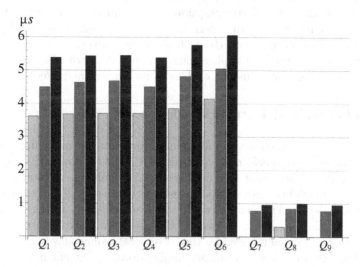

Fig. 7. Query performances in microseconds (logarithmic scale) for each query: black bars refer to 10^5, dark gray to 10^6, and light gray to 10^7 data size.

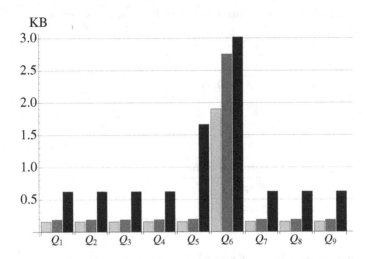

Fig. 8. Main memory consumption in kilobytes for each query: black bars refer to 10^5, dark gray to 10^6, and light gray to 10^7 data size

generated so far. Fig. 7 shows the average query performance times in microseconds (μs) -logarithmic scale- for each query with respect to 10^5, 10^6 and 10^7 stay records. T executes such queries as tensor applications: the monitoring queries (*i.e.* Q1-Q5) has to traverse the sparse matrix, location-by-location, while the path consistency query (*i.e.* Q6) has to traverse the matrix for each EPC, that is the most complex operation. The most trivial operations are the cleaning (*i.e.* Q7-Q9). On average, T performs in the range $[0.006, 87]\,msec$.

Also in this case, another strong point of T is a very low consumption of memory, due to the *sparse matrix* representation of tensors and vectors. Fig. 8 illustrates the main memory consumption of each query with respect to 10^5, 10^6, and 10^7 stay records. On the average, the queries require very few bytes of memory for any dataset, *i.e.* few KBytes. Results demonstrate how our approach can be used in a wide range of applications, where devices with limited calculus resources may process large amount of data in an efficient and effective way.

9   Conclusion and Future Work

This manuscript illustrated a novel algebraic framework for modeling a supply chain, and for efficiently performing analysis of RFID. Our approach proposes a general theoretical model that leverages tensorial calculus, manifesting a great simplicity, and flexibility with multidimensional queries at diverse granularity and complexity levels. The main advantages of our method are the theoretical soundness, and the possibility to exploit the underlying computational hardware. In fact, tensor calculus provides the theoretical grounds for both modeling and querying, and due to its ubiquitous applications, we benefit from low-level optimized code tailored for linear and multilinear (sparse) algebraic numerical manipulation. Experimental results proved our method efficient when compared to recent approaches, yielding the requested outcomes in memory constrained architectures. Our approach can execute queries with few Kilobytes of main memory, and in the case of decentralized query execution, the overall time required by our system do not exceed 100 milliseconds. Massive RFID data analyses show that our framework is 5 to 100 times faster on average, when compared to other modern approaches.

Future investigations are currently under development. We are exploring the introduction of reasoning capabilities, along with a thorough deployment in highly distributed Grid environments. In addition, we are about to test our model on highly memory constrained environment. In particular, we are utilizing recent mobile devices, such as smartphones, and browsers with a full AJAX client-side application.

References

1. Smith, A.: Exploring radio frequency identification technology and its impact on business systems. Information Management and Computer Security 13(1), 16–28 (2005)
2. Tajima, M.: Strategic value of RFID in supply chain management. Journal of Purchasing & Supply Management (2007)
3. EPC Global, http://www.epcglobalinc.org/home
4. Angeles, R.: RFID technologies: supply-chain applications and implementation issues. Information Systems Management 22(1), 51–65 (2005)
5. Bai, Y., Wang, F., Liu, P., Zaniolo, C., Liu, S.: RFID data processing with a data stream query language. In: Proc. of the 23rd International Conference on Data Engineering, ICDE 2007 (2007)
6. Gonzalez, H., Han, J., Li, X., Klabjan, D.: Warehousing and analyzing massive RFID data sets. In: Proc. of the 22nd Int. Conf. on Data Engineering, ICDE, p. 83 (2006)
7. Lee, C.H., Chung, C.W.: Efficient storage scheme and query processing for supply chain management using RFID. In: Proc. of the Int. Conf. on Management of Data, SIGMOD, pp. 291–302 (2008)
8. Papadimitriou, S., Sun, J., Faloutsos, C.: Streaming pattern discovery in multiple time-series. In: Proc. of the 31st Int. Conf. on Very Large Data Bases, VLDB, pp. 697–708 (2005)
9. Bai, Y., Wang, F., Liu, P., Zaniolo, C., Liu, S.: RFID data processing with a data stream query language. In: Proc. of Int. Conf. on Data Engineering, ICDE, pp. 1184–1193 (2007)
10. Jeffery, S.R., Garofalakis, M.N., Franklin, M.J.: Adaptive cleaning for RFID data streams. In: Proc. of the 32nd Int. Conf. on Very Large Data Bases, VLDB, pp. 163–174 (2006)
11. Jeffery, S.R., Alonso, G., Franklin, M.J., Hong, W., Widom, J.: A pipelined framework for online cleaning of sensor data streams. In: Proc. of the 22nd Int. Conf. on Data Engineering, ICDE, p. 140 (2006)
12. Wang, F., Liu, S., Liu, P., Bai, Y.: Bridging Physical and Virtual Worlds: Complex Event Processing for RFID Data Streams. In: Ioannidis, Y., Scholl, M.H., Schmidt, J.W., Matthes, F., Hatzopoulos, M., Böhm, K., Kemper, A., Grust, T., Böhm, C. (eds.) EDBT 2006. LNCS, vol. 3896, pp. 588–607. Springer, Heidelberg (2006)

13. Hoffman, M., Kunze, R.: Linear Algebra. Prentice Hall (1971)
14. Abraham, R., Marsden, J.E., Ratiu, T.: Manifolds, Tensor Analysis, and Applications. Springer (1988)
15. Heinbockel, J.H.: Introduction to Tensor Calculus and Continuum Mechanics. Trafford Publishing (2001)
16. Jancewicz, B.: The extended grassmann algebra of $\mathbb{R}^3$. In: Clifford (Geometric) Algebras with Applications to Physics, Mathematics, and Engineering. Birkhäuser, Boston (1996)
17. Derakhshan, R., Orlowska, M.E., Li, X.: RFID data management: Challenges and opportunities. In: Proc. of the IEEE Int. Conf. on RFID, pp. 175–182 (2007)
18. Bondy, A., Murty, U.S.R.: Graph Theory. Graduate Texts in Mathematics. Springer (2010)
19. Chartrand, G.: Introductory Graph Theory. Dover Publications (1984)
20. Duff, I.S., Erisman, A.M., Reid, J.K.: Direct Methods for Sparse Matrices. Numerical Mathematics and Scientific Computation. Oxford Univ. Press (1989)
21. Davis, T.A.: Direct Methods for Sparse Linear Systems. SIAM (2006)
22. Osterby, O., Zlatev, Z.: Direct Methods for Sparse Matrices. LNCS, vol. 157. Springer, Heidelberg (1983)
23. Sears, M.P., Bader, B.W., Kolda, T.G.: Parallel implementation of tensor decompositions for large data analysis. In: SIAM AN 2009. SIAM (July 2009)
24. Lin, C.Y., Liu, J.S., Chung, Y.C.: Efficient representation scheme for multidimensional array operations. IEEE Transactions on Computers 51, 327–345 (2002)
25. Lin, C.Y., Chung, Y.C., Liu, J.S.: Efficient data compression methods for multidimensional sparse array operations based on the ekmr scheme. IEEE Trans. Comput. 52, 1640–1646 (2003)
26. Bader, B.W., Kolda, T.G.: Algorithm 862: Matlab tensor classes for fast algorithm prototyping. ACM Transactions on Mathematical Software 32(4), 635–653 (2006)
27. Bader, B.W., Kolda, T.G.: Efficient MATLAB computations with sparse and factored tensors. SIAM Journal on Scientific Computing 30(1), 205–231 (2007)
28. Bertsekas, D.P., Tsitsiklis, J.N.: Parallel and Distributed Computation: Numerical Methods. Athena Scientific (1997)
29. Williams, S., Oliker, L., Vuduc, R., Shalf, J., Yelick, K., Demmel, J.: Optimization of sparse matrix-vector multiplication on emerging multicore platforms. In: Proc. SC 2007: High Performance Computing, Networking, and Storage Conference, pp. 10–16 (2007)
30. Johnson, R.W., Huang, C.H., Johnson, J.R.: Multilinear algebra and parallel programming. In: Proceedings of the 1990 ACM/IEEE Conference on Supercomputing, Supercomputing 1990, pp. 20–31. IEEE Computer Society Press, Los Alamitos (1990)
31. Buluç, A., Fineman, J.T., Frigo, M., Gilbert, J.R., Leiserson, C.E.: Parallel sparse matrix-vector and matrix-transpose-vector multiplication using compressed sparse blocks. In: Proc. of the 21st Annual Symposium on Parallelism in Algorithms and Architectures, SPAA 2009, pp. 233–244. ACM (2009)
32. Iverson, K.: A programming language. In: Proc. of the AFIPS Spring Joint Computer Conference (1962)
33. Kleene, S.C.: Introduction to Metamathematics. Van Nostrand Rheinhold (1952)

Processing Exact Results for Windowed Stream Joins in a Memory-Limited System: A Disk-Based, Adaptive Approach

Abhirup Chakraborty[1,*] and Ajit Singh[2]

[1] Department of Informatics and Computing
Indiana University, Bloomington, IN 47408, USA
[2] Dept. of Electrical and Computer Engineering
University of Waterloo, ON, Canada, N2L3G1
achakrab@indiana.edu, asingh@uwaterloo.ca

Abstract. We consider the problem of processing exact results for sliding window joins over data streams with limited memory. Existing approaches either, (1) deal with memory limitations by shedding loads, and therefore cannot provide exact or even highly accurate results for sliding window joins over data streams showing time-varying rate of data arrivals, or (2) suffer from large I/O overhead due to random disk flushes and disk-to-disk stages with a stream join, making the approaches inefficient to handle sliding window joins. We provide an Adaptive, Hash-partitioned Exact Window Join (AH-EWJ) algorithm incorporating disk storage as an archive. Our algorithm spills window data onto the disk on a periodic basis, refines the output result by properly retrieving the disk-resident data, maximizes output rate by employing techniques to manage the memory blocks, and continuously adjusting the allocated memory within the stream windows. The problem of managing the window blocks in memory—similar in nature to the caching issue—captures both the temporal and frequency related properties of the stream arrivals. We present a baseline algorithm called Rate-based Progressive Window Joins (RPWJ), which extends an existing algorithm to tune the performance by reducing disk I/O overhead while processing sliding window joins. We provide experimental results demonstrating the performance and effectiveness of the proposed algorithm.

1 Introduction

With the advances in technology, various data sources (sensors, RFID Readers, web servers, etc.) generate data as high-speed data streams. Traditional DBMSs, that process streams of queries over a non-streaming database, are not suitable for processing data streams where long running continuous queries over bursty, high-volume online data need to be processed in online fashion to generate results in real time. Thus the new requirements unique to the data stream processing systems pose new challenges

---

* The research work was done while the first author was a Ph.D. Candidate at the University of Waterloo, Canada.

A. Hameurlain et al. (Eds.): TLDKS VII, LNCS 7720, pp. 31–61, 2012.

providing the motivation to develop efficient techniques for data stream processing systems. Examples of applications relevant to data stream processing systems include network traffic monitoring, fraud detection, financial monitoring, sensor data processing, etc. Considering the mismatch between the traditional DBMS and the requirements of data stream processing, a number of *Data Stream Management Systems* (DSMS) have emerged [1–3]. These systems aim at processing time-varying data streams in real time. In this paper we investigate the problems that arise when processing joins over data streams.

Computing the approximate results based on load shedding [4–7] is not feasible for queries with large states (e.g., join with large window size) as the accuracy of the query output might be below the QoS [1] specified by the user. This situation can easily arise in a scenario where the state size is far larger than the allocated memory, and the majority of the tuples, including the productive ones, should be dropped. It is formally shown in reference [5] that given a limited memory for a sliding window join, no online strategy based on load shedding can be k-competitive for any k that is independent of the sequence of tuples within the streams. Thus, for a system with QoS-based query output, secondary storage is necessary even to guarantee a QoS above a certain limit. Moreover, there exist many application scenarios where continuous sliding window join queries in the system should produce exact results even though the query system may not have enough memory to cope up with the query workload at runtime [6, 8]. For example, applications like decision support, intelligence or disaster monitoring may run the risk of "missing the needle in the haystack" when provisioned with load-shedding technique [6].

Pushing the operator states or stream tuples to the disk has already been adopted in several research works [9–11] that aim at providing high output rate and complete query results for a single operator query over finite data sets. Reference [8] proposes mechanisms to push states onto disks for queries with multiple operators. However, in all of the above strategies, that process finite data sets, the clean-up phase occurs at the end when there does not exist any stream tuples for processing. Contrary to this scenario, in the case of a sliding window stream join with bursty input arrival patterns, the in-memory execution and disk clean-up phases should be interleaved, and thus be scheduled properly.

Algorithms for processing joins over finite streams [9, 12, 10] also suffer from the problem as stated above. Moreover, these algorithms are not I/O-efficient and do not consider *disk I/O amortization* over large number of input tuples: operations involving flushing memory tuples and joining disk resident tuples are invoked at a partition level, and such operations require a large number of small random disk I/Os. Also, these algorithms result in a low memory utilization and a high overhead of eliminating duplicate tuples in join output. Thus exact processing of sliding window stream joins within a memory limited environment is a significant and non-trivial research issue as promulgated in [6, 8, 5].

In this paper, we consider the issue of processing exact results for sliding window joins over data streams. Using disk storage as an archive, we propose an Adaptive, Hash Partitioned Exact Window Join (AH-EWJ) algorithm that endeavors to smooth the load by spilling a portion of both the window blocks onto the disk. We propose a framework for processing disk-based sliding window joins. The proposed algorithm merges the disk-resident data periodically with the in-memory blocks during a phase called *disk probing*. This approach merges the flushed tuples onto one disk partition and amortizes a disk-access over a large number of input tuples, thus eliminating small, random disk I/O; it improves memory utilization by employing *passive removal* of the blocks from the stream window. Also, this proposed algorithm dynamically adjusts the stream windows and fine tunes the hash buckets within each stream (during *passive removal*) based on the stream arrival patterns. To increase the output generation rate, AH-EWJ algorithm employs a generalized framework to manage the memory blocks forgoing any assumption about the models (unlike previous works e.g. [5]) of stream arrival. In summary, the key contributions of the paper are as follows:

1. We propose a disk-based join algorithm AH-EWJ that, given a limited memory, processes sliding window joins over arbitrary sequences of streaming data. The proposed algorithm produces correct output by spilling data onto the disk and re-trieving/joining the spilled data within the time bound set by the window.
2. We propose a disk probing policy that, on one hand, maximizes disk efficiency by amortizing a disk scan over a number of tuples. On the other hand, the same policy maintains the productive blocks within memory, thus maximizing output rate. The proposed technique can cope with arbitrary patterns of stream arrivals.
3. We propose a methodology to adjust both the window and the partition sizes based on the arrival rates of the streams. Our proposed scheme avoids disk dump when-ever the windows can be incorporated within the memory.
4. We present detailed experimental studies showing the effectiveness of our tech-niques.

The rest of the paper is organized as follows. Section 2 provides a survey enumerating the research works related to ours. Section 3 provides the basic concept in processing sliding window queries. Section 4 defines the problem and provides an overview of the proposed algorithm. Section 5 describes the techniques to manage memory blocks, the procedure to integrate periodically the disk and memory blocks (disk probing), and finally the AH-EWJ algorithm. Section 7 describes experimental methodologies in de-tails, and presents the experimental results comparing the performance of the proposed algorithm (AH-EWJ) with that of the baseline one (RPWJ). Section 8 concludes the paper and outlines the potential future works.

2   Related Works

Existing join algorithms on streaming data can be classified into two categories: the first one considers bounded or finite size relations, whereas the second category considers the streams of infinite size. The *finite stream joins* focus on generating progressive re-sults. In case of the second category, the tuples are joined based on sliding windows that span a time interval in recent past of the data streams.

2.1 Join over Bounded Stream

There exist numerous join algorithms for joining streaming finite data sets in a non-blocking fashion. Symmetric Hash Join [13], that extends the traditional hash join algorithm, is the first non-blocking algorithm to support pipelining. The XJoin algorithm [9] rectifies the situation by incorporating disk storage in processing joins: when memory gets filled, the largest hash buckets among A and B is flushed into disk. When any of the sources is blocked, XJoin uses the disk resident buckets in processing join. In reference [11], the authors present multi-way join (MJoin) operators, and claims performance gain while compared with any tree of binary join operators. Progressive-Merge Join (PMJ) algorithm [14] is the non-blocking version of the traditional sort-merge join. The Hash-Merge-Join (HMJ) [10] algorithm combines advantages of XJoin and PMJ. The flushing policy aims at maximizing the output generation rate by balancing total in-memory tuples from each stream. RPJ (rate-based progressive join) algorithm focuses on binary joins, and extends the existing techniques (i.e., those proposed in XJioin, HMJ or MJoin) for progressively joining stream relations. Reference [8] considers queries with multiple operators and proposes state spill mechanisms in order to maximize the output rate. The spilled state or data are joined at the end during a clean-up phase.

In [15], the authors propose a join operator, that reduces the production delay of the result tuples by producing join results early. This algorithm is based on a state manager that switches the join processing between in-memory data and disk-resident data in order to maximize overall throughput. RIDER [16] processes joins using a Nested-Loop-Join algorithm. The primary objective of the algorithm is to (a) devise a flushing policy to maximize output rate, and (b) enable the system quickly switch between a in-memory stage and disk-to-disk stage. Unlike the RPJ algorithm, the RIDER algorithm does not have any disk-to-memory stage.

All these algorithms are applicable in case of joining streams with finite number of tuples. These algorithms invokes reactive stages (i.e., join disk-resident data with in-memory data or disk-resident data) when the CPU is idle, and the clean-up process occurs at the end. However, in case of the sliding window joins over unbounded streams, the clean up or invalidation is a continuous process and should be interleaved with the stream processing.

2.2 Join over Unbounded Data Stream

Processing joins over infinite stream usually employs a certain windowing technique to limit the memory requirements of continuous queries, thus unblocking query operators. There has been intense research works on processing stream joins over sliding windows. Here, we summarize the significant works on this issue. It should be noted that all of the algorithms shed loads during bursty stream arrivals and are not accurate.

The work presented in [17] investigates the techniques to process sliding window joins over a pair of unbounded streams. The authors introduce a unit-time-basis cost model to analyze the performance of the join techniques. The cost model consists of two separate terms each one corresponding to a join direction. Decoupling the cost of each join direction, the proposed technique demonstrates the effectiveness of an asymmetric join technique, where a different technique is used for different stream to be joined (e.g., nested loop join for one direction and hash join for the other). Although the main focus of the paper is not load shedding, it also considers the scenarios where system resources are not sufficient to keep up with the input streams.

Reference [6] considers the problem of approximate join evaluation for a pair of data streams. The paper examines the MAX-subset measure and presents optimal offline algorithms for sliding window joins. It also presents some heuristics for the online join processing. In [18], the authors consider a different model by incorporating *importance* semantics within input tuples and the presents optimal offline algorithms and online heuristics that seek to maximize the importance of the approximate join result.

Golab et. al. [19] presented and evaluated various algorithms for processing multi-joins over sliding windows residing in main memory. Here, multi-joins are evaluated together in a series of nested for-loops, and newly arrived tuples from each window are processed separately (possibly using different join orders). Based on a per unit-time cost model, the authors propose join ordering heuristics that try to lower the number of intermediate result tuples passed down to the inner loops.

Srivastava et. al [5] propose algorithms for performing memory-limited stream joins. The age-based model for stream arrival introduced in the paper captures many applications not captured by the frequency-based model. This model is based on the time correlation between the streams. Specifically, the probability of a match between a pair of tuples residing on a pair of streams depends on the difference between their times-tamps. Based on this observation, the proposed algorithm conserves memory by keeping an incoming tuple in a window up to the point in time until the average rate of output tuples generated using this tuple reaches its maximum value. In [4], the authors pro-pose an adaptive CPU load-shedding approach for multi-way windowed stream joins. Instead of processing the whole window for each incoming tuple, this approach selec-tively processes a subset of the join window through three types of run time adaptations: adaptation to input stream rates, adaptation to time correlation between the streams and adaptation to join directions. This approach (1) captures time based correlation between streams by maintaining statistics at segment level, and then (2) revises the selective pro-cessing decisions based on the statistics. In [20], the authors present a window-based stream joins called Handshake-joins that exploit parallelism in multi-core processors. Reference [21] joins an unbounded stream with a persistent relation (stored on the disk) considering the non-uniform access costs of the input tuples; the join algorithm mini-mizes disk overhead using the notion of *disk I/O amortization*. A preliminary version of this work appeared in [22].

3   Preliminaries

We briefly describe the basic model of processing continuous, sliding window queries over data streams. *Sliding windows* [3] are used to limit the state size in a stateful operator. For a stream $S_i, i = 1, 2$, we use r_i to denote the average arrival rate in stream S_i. In a dynamic stream environment, this arrival rate can change over time. Each tuple $s \in S_i$ has a timestamp $s.t$ identifying the arrival time at the system. As in [23], we assume that the tuples within a stream have a global ordering based on the system's clock. We use $S_i[W_i]$ to denote a sliding window on the stream S_i, where W_i is the window size in time units. At any time t, a tuple s belongs to $S_i[W_i]$ if s has arrived on S_i within the interval $[t - W_i, t]$. An arriving tuple s in stream S_i is processed as shown in Figure 1 (for notational convenience $S_{\bar{i}}$ is denoted to be the opposite stream).

1: When a new tuple s arrives in stream s_i
2:      Invalidate: Discard expired tuples in $S_i[W_i]$
3:      Probe: Generate $s \bowtie S_{\bar{i}}$
4:      Insert: Add tuple s to $S_i[W_i]$

Fig. 1. Processing a newly arrived tuple in Sliding Window Joins

4   Problem Definition and Solution Overview

The basic join operator considered in this paper is a sliding window equi-join between two streams S_1 and S_2 over a common attribute A, denoted as $S_1[W_1] \bowtie S_2[W_2]$. The output of the join consists of all pairs of tuples $s_1 \in S_1, s_2 \in S_2$ such that $s_1.A = s_2.A$, and $s_1 \in S_1[W_1]$ at time $s_2.t$ (i.e., $s_1.t \in [s_2.t - W_1, s_2.t]$) or $s_2 \in S_2[W_2]$ at time $s_1.t$ (i.e., $s_2.t \in [s_1.t - W_2, s_1.t]$. We assume that the amount of memory available for storing the join operator states or windows is less than the size of the windows. Hence, a portion of each of the stream window W_i resides on the disk. The proposed algorithm (AH-EWJ) is based on the hashing methodology. Tuples in the stream windows are mapped to one of the n_{part} partitions, using a hash function $\mathcal{H}$ that generates an integer in the range of $[1, n_{part}]$. Hence, the join algorithm should keep a portion of each of the hash partitions (or , buckets) into the memory and spill the remaining portion onto the disk. The join operator has buffers attached with its input streams. The join operator fetches tuples from the input buffers, processes the join, and sends the output as a stream; it must ensure that each result tuple is generated exactly once.

The straightforward approach to the sliding window join over unbounded streams is to use the existing algorithm XJoin [9] (or, its variants, i.e.,RPJ [12]), and invoke the clean up or reactive stage (i.e., disk-to-memory and disk-to-disk) periodically. But, this would result in significant processing and disk overhead as the scan of a disk or memory segment might occur for a few tuples within a partition. In this approach, the number of disk I/Os for both the invalidation (of the stream windows) and the disk dump operation would also be $O(n_{part})$.

In addition to the above limitations, our approach is motivated by the following observations on the performance of XJoin and its variants. Firstly, these approaches attains low utilization of memory. When, the memory is full, XJoin flushes tuples (N_{flush}) from a few selected partitions. Hence, on the average, a partition of memory equivalent to the size of $\frac{N_{flush}}{2}$ tuples remains unutilized. This unutilized memory can be decreased by selecting a low N_{flush}, but this in turn increases the I/O overhead as it does not amortize a disk write over a large number of input tuples. Hence, the proposed approach should eliminate such holes in the memory, and make the disk flush efficient by eliminating frequent and small disk writes. Also, the duplicate elimination is a complex process with significant processing and storage overhead.

We propose an algorithm called *Adaptive, Hash Partitioned Exact Window Join* (AH-EWJ) that incorporates the above observations. The algorithm computes sliding window joins between two streams without compromising the output accuracy. Existing algorithms shed the load by dropping tuples selectively. Contrary to these algorithms, we attempt to defer the load during high workload, and process the deferred load during the period of low workload. Thus, AH-EWJ aims at smoothing the load using the disk as secondary storage. The proposed algorithm AH-EWJ is based on the framework given in Figure 2 showing the organization of tuples within a stream window W_i. For each stream $S_i(i = 1, 2)$, AH-EWJ stores the relevant tuples (window states) in the memory and on the disk. We denote the portion of the window $S_i[W_i]$ residing in memory and disk as W_i^{mem} and W_i^{disk} respectively. We assume that the sizes of a memory page and a disk block are equal, and use the terms interchangeably. We arrange the tuples into pages (or blocks) and process the tuples at the granularity of a page or block. In addition to the timestamp of the tuples within a page, each page or block contains a unique sequence number (or timestamp). All tuples within a block are homogeneous in the sense that they are probed against the same number of blocks from the window of the opposite stream $(S_{\bar{i}}[W_{\bar{i}}])$.

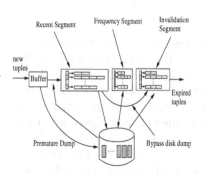

Fig. 2. Organization of incoming tuples within a window

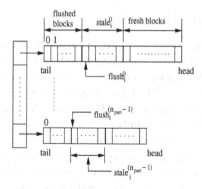

Fig. 3. Data structure of a Recent Segment of a Stream Window

AH-EWJ eliminates a large number of small, random I/Os (one for each partition) by merging the flushed blocks into one disk segment and performing disk read on this unified segment. The memory segment of a window W_i is divided into an *invalidation segment* and a *generative segment*. An invalidation segment is allocated to reduce disk I/O while expiring the tuples from the stream window which makes the tuple expiration efficient; the generative segment stores the more productive blocks to maximize the output throughput. Generative segment has two parts: *frequent segment* and *recent segment*. An incoming tuple s in stream S_i is joined with the memory resident portion of the hash partition from the opposite window (i.e., $W_{\bar{i}}^{mem}[\mathcal{H}(s)]$). Upon memory overflow, the tuples from the recent segment are dumped onto the disk during a phase called

Disk Probing. The algorithm updates the frequent segment periodically at the elapse of an interval. The algorithm adapts memory allocation in two levels: at window level, it adjust the window size based on arrival rates, and within a stream window, it adjusts the partition sizes by evicting blocks from a partition with low productivity, that records the number of output tuples generated by the tuples in a block.

5 Exact Join

In this section, we describe in details the join algorithm for processing sliding window join for streaming data. The algorithm consists of four major sub-tasks: invalidation of the tuples, maintaining blocks in generative segments, adapting the memory allocation both within and across the stream windows, and probing the disk periodically to join the disk-resident data with the incoming data. Before presenting the algorithm we illustrate the key ideas encompassing the major subtasks. The generative segment of a window W_i is divided into two segments: Recent segment (W_i^{rec}) and Frequent segment (W_i^{freq}). So, the memory portion of each window W_i has three segments in total: invalidation segment (W_i^{inv}), Recent segment and Frequent segment. Each of the segments is maintained as a hashtable with n_{part} partitions or buckets. We denote the j-th partition of a segment W_i^{seg} as $W_i^{seg}[j]$, where $seg \in \{rec, freq, inv, mem\}$.

5.1 Memory Stage

The incoming tuples in stream S_i are mapped to the respective partitions using a hash function $\mathcal{H}$ and added to the head block of the partition in the recent segment $W_i^{rec}[p][H]$. Each newly added block within a stream is given a unique number denoted as *block number*. Other than the block number, every block is associated with a *productivity value* that records the number of output tuples generated by the tuples in that block. The block number is used in eliminating redundant output tuples as described in the subsequent part of this section. The arriving tuples are joined with the respective partition in the memory segment of the opposite window $W_{\bar{i}}^{mem}[p]$. Blocks (in stream S_i) are added to the recent segment at one end (i.e., head), while the blocks from the other end (i.e., tail) are stored on the disk during a disk dump. In our join algorithm, we maintain the following constraint:

> **Constraint 1**: *At a particular time, the blocks from recent segment of stream S_i should join with all the blocks in the disk portion of window $W_{\bar{i}}$ (i.e., $W_{\bar{i}}^{disk}$) before being dumped to the disk (during mature dump)*

Blocks in the recent segment are categorized into three types: flushed blocks, stale blocks and fresh blocks. The fresh blocks are newly arrived blocks that haven't yet participated in the disk probing. The stale blocks have already participated in the disk probing, but have not been flushed onto the disk. The flushed blocks have already been dumped onto the disk, and, therefore, can be deallocated to accommodate newly arrived blocks. During the disk probing phase as described in subsection 5.2, blocks retrieved from the disk segment (of the opposite stream) are joined with the *fresh blocks* in the recent segment. After the disk probing, the fresh blocks are marked as stale ones. Blocks from the recent segment are evicted upon the arrival of new tuples, and the victim blocks are selected adaptively based on both the stream arrival patterns and productivity values (described in subsection 5.5). A victim partition within a recent segment usually evicts a flushed block. However, as described in the subsection 5.5 however, free blocks from the invalidation segment can sometimes be evicted. Figure 3 shows the structure of a stream window. Here, $flush_i^p$ is the location of the first stale block in a partition p of recent segment of stream S_i, and $stale_i^p$ is the number of stale blocks in the same partition. Blocks in the locations preceding $flush_i^p$ are the flushed blocks; those in the range $[flush^p, flush_i^p + stale_i^p - 1]$ are the stale blocks; and those in the range $[flush_i^p + stale_i^p, |W_i^{rec}[p] - 1|]$ are the flesh blocks. After disk probing, the value of $stale_i^p$ increases; whereas, after a disk dump, the value of $stale_i^p$ decreases and that of $flush_i^p$ increases, setting the periphery of the flushed blocks.

It should be clear that a disk dump in stream S_i merges the N_{flush} number of stale blocks from all the partitions in the recent segment W_i^{rec} and flushes the tuples onto the disk. During a disk dump, blocks from a partition are stored sequentially, that minimizes partition switches while joining the blocks with the respective partition during the disk probe (we omit the detailed technique and refer the readers to [21]). Note that, the blocks dumped onto the disk (i.e., flushed blocks) are kept in the partitions, and these blocks are joined with the blocks/tuples arriving in the opposite stream. The flushed blocks are removed only on demand. Such lazy removal of the flushed blocks achieves high memory utilization increasing the output generation rate. Without lazy removal, the recent segment might fluctuate between overflow and empty states: on the average, half of the recent segment would remain unutilized.

5.2   Disk Probing

In this subsection, we present the basic techniques and data structures involved with disk probing.

Frequent Segment Update. The decision about placement or replacement of a block in Frequent segment is based on its productivity value. This decision is made periodically after a certain interval. The productivity values of the blocks are decayed during this update stage. The feasible time to carry out this update step is during the disk probing or mature dump when the all the disk blocks are scanned. During the update stage, as the disk blocks are scanned, a block is brought into the Frequent segment if its productivity exceeds that of a block already in Frequent segment at least by a fraction (ρ). The block having the minimum productivity value among the blocks in the Frequent segment is replaced.

Productivity Table. All the blocks within a window need to store the productivity values. Updating this value for a memory resident block is easy; however, updating the productivity value for a disk resident block requires one disk write to update the corresponding block. To make this update of productivity values efficient we maintain the productivity values of all the disk resident blocks in memory as a list of tuples <*block number, productivity*>, and use the term productivity table to refer to this list. Such a table consumes little space and might be stored in the memory. The productivity values of the blocks in a Disk segment are updated during disk probing.

Eliminating Redundant Tuples. A block b_i^p evicted from the Frequent segment of window W_i is already joined with the fresh blocks in the Recent segment of the stream $W_{\bar{i}}$. If the evicted block is not already scanned on the disk, it will be read and be joined with the fresh blocks in the Recent segment of window $W_{\bar{i}}$. To prevent this duplicate processing, we maintain a list E_i containing the block numbers of evicted blocks. If an incoming block from the disk is contained in E_i, we omit processing that block. List E_i is reset to *null* at the end of the update stage. The productivity values of all the blocks in the disk segment and the frequent segment are decayed (using a factor θ $(0 < \theta < 1)$) after the update stage.

As described earlier, blocks from the Recent segment are removed on demand. Such passive removal of the blocks might lead to duplicate output generation. Let us consider a block b_i^j in partition $\mathcal{H}(b_j^i)$ of the recent segment W_i^{rec}. The block b_i^j joins with a block $b_{\bar{i}}^k \in W_{\bar{i}}^{rec}[\mathcal{H}(b_j^i)]$. Later b_i^j participates in mature dump and is stored on the disk. However, due to the passive removal of the blocks from the Recent segment, the same block (b_i^j) remains in memory as a stale block. Now, when block $b_{\bar{i}}^k$ of the opposite window participates in mature dump, it finds on the disk the block b_i^j already joined in previous step. We use the sequence number of a block, denoted as *block number*, in solving this issue: every incoming block is assigned an unique number from an increasing sequence. Every block b_i^k in partition $\mathcal{H}(b_i^k)$ of the Recent segment W_i^{rec} stores the minimum block number (*minBN*) among the blocks, from the same partition of the Recent segment of the opposite window, that b_i^k joins with. When b_i^k participates in mature dump, it joins with a block $b_{\bar{i}}^p \in W_{\bar{i}}^{rec}[\mathcal{H}(b_k^i)]$ if the block number of the block $b_{\bar{i}}^p$ is less than the *minBN* value of the block b_i^k, i.e., $b_{\bar{i}}^p.BN < b_i^k.minBN$. As b_i^k is already in W_i^{mem}, any block in $W_{\bar{i}}$ arriving after $b_i^k.minBN$ is already joined with b_i^k. So, during disk probing any block $b_{\bar{i}}^p \in W_{\bar{i}}^{disk}$ having $b_{\bar{i}}^p.BN \geq b_i^k.minBN$ can be omitted.

On the other hand, we observe that any block b_i^p in Frequent segment of window W_i (i.e., W_i^{freq}) is already joined with the corresponding partition in the memory segment of window $W_{\bar{i}}$ (i.e., $W_{\bar{i}}^{rec}$, $W_{\bar{i}}^{inv}$). When a block is brought into the Frequent segment, it is already joined with the blocks in disk segment of window $W_{\bar{i}}$, i.e, $W_{\bar{i}}^{disk}$ (refer to **Constraint 1** stated earlier), and that block is joined with $W_{\bar{i}}^{rec}$ immediately before being placed on the Frequent segment. Now, any incoming block in window $W_{\bar{i}}$ is joined with b_i^p. Moreover, if any block is dumped on the disk, it has already found b_i^p on memory and has been joined with that block. No issue of duplicate processing other than the one mentioned above arises with frequent segment. Hence, blocks in frequent segment need not store any extra information.

Probing Algorithm. Procedure DISKPROBE(), as shown in Procedure 1, is invoked periodically when the number of fresh blocks within a Recent segment exceeds a predefined threshold N_{rec}^{min} (cf. 5.5). The procedure scans the disk retrieving blocks in $W_{\bar{i}}^{disk}$ and joins the blocks with those in W_i^{rec}. Blocks in the Frequent segment of window $W_{\bar{i}}$ (i.e., $W_{\bar{i}}^{freq}$) are updated periodically, and this update phase is carried out with disk probing. Moreover, this *update interval* might be longer than the *probing interval*—the interval between two consecutive disk probes in a stream. So, the parameter $uFlag$ is used to indicate when to update the relevant Frequent segment(i.e., $W_{\bar{i}}^{freq}$).

Procedure 1. DISKPROBE(W_i^{rec}, W_i^{disk}, W_i^{freq}, $uFlag$)

1: position disk head to the first block in W_i^{disk}
2: $E_i \leftarrow null$
3: **while** exists new block in W_i^{disk} **do**
4:    read block b_i^m from disk
5:    $p \leftarrow \mathcal{H}(b_i^m)$
6:    **for** $n \leftarrow (flush_i^p + stale_i^p)$ to $|W_i^{rec}[p]| - 1$ **do**
7:      **if** $b_i^m.BN < b_i^n.minBN$ **and;** $b_i^m \notin W_i^{freq}[p]$; **or** $b_i^m \notin E_i$ **then**
8:       join b_i^n with b_i^m
9:       update $b_i^m.prod$ and $b_i^n.prod$
10:       $b_i^n.minBN \leftarrow null$
11:       **if** $uFlag$=TRUE **then**
12:        $bn \leftarrow$ UPDATEFREQSEGMENT(W_i^{freq}, b_i^m)
13:        **if** $bn \geq 0$ **then**
14:         insert bn into E_i
15:        **end if**
16:       **end if**
17:      **end if**
18:    **end for**
19: **end while**

Procedure 2. UPDATEFREQSEGMENT(W_j^{freq}, b_j^m)

Require: return: evicted block number
       local variables: bn, p
1: $b_{min} \leftarrow 0$ MINPRODBLOCK(W_j^{freq})
2: $bn \leftarrow -1$ {a negative number}
3: $p \leftarrow \mathcal{H}(b_j^m)$
4: **if** $b_j^m.prod > (1 + \rho)b_{min}$ **then**
5:    $bn \leftarrow b_{min}.BN$
6:    Evict b_{min} from $W_i^{freq}[\mathcal{H}(b_{min})]$
7:    copy b_i^m to b_{min}
8:    insert block b_i^m into $W_i^{freq}[p]$
9: **end if**
10: **return** bn

To eliminate duplicate tuples in output, an incoming block from the disk is checked against the condition in line 7 before joining with the fresh blocks in W_i^{rec}. The first part of the condition, as described earlier in this section, ensures that the disk block was not joined with the memory resident block (from W_i^{rec}) before being dumped onto the disk, while the second part checks if the incoming block is already in the Frequent segment or is recently evicted from the Frequent segment. In line 9, the productivity values of the blocks are updated on the productivity table avoiding expensive disk writes. Line 12 invokes the procedure UPDATEFREQSEGMENT, which updates the Frequent segment W_i^{freq}. If the input block is stored within the frequent segment, the procedure returns the non-negative block number of the evicted block.

5.3 Invalidation

As the window slides over time, tuples from the tail of the window should be discarded continuously. To reduce the disk access time in eliminating tuples, we reserve a few blocks in memory to store the tail of the window and term the blocks as *invalidation segment*. Without this invalidation segment (W_i^{inv}), in the worst case, we need to access the disk for each incoming tuple. Using this invalidation segment, we discard the tuples from the memory blocks whenever the window slides over time; the disk segment is accessed only when the invalidation segment of the corresponding window can accommodate a chunk of blocks B^{win} from the disk segment termed as Basic Window.

5.4   Disk Dump

As illustrated earlier, during disk dump, blocks from all the partitions of a recent segment are merged and stored on the disk as a unified disk segment of the corresponding window. Now, if the blocks in the disk segment are not sequenced according to their time-stamps, the process of the invalidation/expiration of tuples might be affected. To realize the scenario, let us consider a disk segment where blocks with tuples of lower timestamps (i.e., older ones) are followed by those with lower timestamps (i.e. more recent ones). Hence, the invalidation segment get filled with the tuples having lower time-stamps (i.e., younger ones). Now, the older tuples can not be brought into memory, and thus invalidated, unless all the younger tuples within the invalidation segment are expired. Such a blocking due to disorder of the blocks within the disk segment unnecessarily halts the expiration of the older tuples, that also delays their processing or tuple generation time. To alleviate the problem, two methods can be used: (1) stale blocks from the recent segment are sequenced according to their block number, (2) blocks from the partitions can be sequenced by fetching one block at a time, from the partitions, in a round robin fashion. However, both of the approaches result in cache thrashing while joining the blocks in the disk segment with the opposite window. This is due to the changes in partitions for each subsequent blocks.

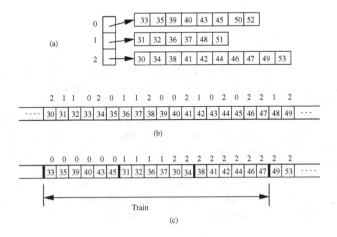

Fig. 4. Mechanism to dump the stale blocks from a recent segment. (a) shows the stale blocks in the hash table. (b) sequence of blocks dumped according to their block number (c) dumping with bounded disorder using the notion of a *train*.

Procedure 3. DISKDUMP(int i, int $N_f lush$)

Require: input: i=stream index; $N_f lush$=total tuples to flush
    return: a freed block b
    global variables: W_i; $i = 0, 1$; $trainSize$, $trainLimit_i$, $dumpC_i$
1: **if** $dumpC_i \geq trainLimit_i$ **then**
2:     $trainLimit_i \leftarrow trainLimit_i + trainSize$
3: **end if**
4: **for** $i \leftarrow 1$ to n **do**
5:     **if** BN of the block at location $flush_i^p$ is equal to $trainLimit_i$ **then**
6:         $p \leftarrow$ next partition with BN of the block at location $flush_i^p$ less than $trainLimit_i$
7:     **end if**
8:     flush the block at location $flush_i^p$ to disk
9:     increment $flush_i^p$ and decrement $stale_i^p$
10:     increment $dumpC_i$
11: **end for**

Based on the above observations, we propose a technique to order the blocks, that stores the blocks from the same partitions in a sequence. Storing the blocks from the same partitions in a succession reduces the cache thrashing (at L1 and/or L2 caches) due to the frequent switching of the partitions while joining the blocks from the disk segment. Figure 4 illustrates the technique to dump the blocks. The blocks are identified by the *block number* BN. All the stale blocks with number less than 30 are either dumped onto the disk to transferred to the invalidation segment (due to an empty disk segment). Now, arranging the blocks according to the block numbers results in large number of partition switches (i.e., successive blocks belong to different partitions) as shown in Figure 4(b). Here, the numbers above the list indicates the partition numbers of the blocks. On the other hand, our approach based on bounded disorder causes only a few partition switches. Here, blocks from the same partition are stored in a sequence. However, storing a large number of tuples from the same partition might lead to disorder, which in turn might block the invalidation segment. We use the notion of a *train size* to bound the disorder within the block numbers. The number of blocks that might be dumped in a sequence is determined by the *train size*. At the start of every *train*, the upper limit of the block number of the train for stream i (i.e., $trainLimit_i$) is determined. Blocks from a partition are dumped until a block with BN higher that upper limit is encountered; in that case, the blocks from the next partition is dumped Figure 4(c) shows the blocks as dumped onto the disk. Here, the train size is 18,and the flush size N_{flush} is 6. As all the lowest BN among the stale blocks is 30 (i.e., 30 might be the upper limit for the previous *train*), the upper limit for the current *train* is 48. Thus each *disk dump* within the *train* arranges blocks from the same partition until no more blocks in the partition is left or the next block in the partition has a BN not less than 48. Note that, during 2nd dump, blocks from the partition 1 are flushed until block 48 is encountered, at which point blocks from the partition 2 are flushed. The boundaries of the blocks within a disk dump are indicated by the bold lines. The partitions are switched in a cyclic fashion once the upper limit is reached. At the end of the 3rd dump, a new train starts with an upper limit 66 (=48+18). Procedure 3 shows the pseudocode for disk dump operation.

5.5 Adapting Window and Bucket Sizes

In the join scheme, sizes of the invalidation and the frequent segments are kept constant; the remnant join memory (M_{free}) is allocated between the recent segments, that change their sizes depending on stream arrival rates. Upon arrival of a new block, if no free blocks are available within the recent segment, the recent segment either evicts a block from itself or receives a block from the opposite window. To ascertain the memory allocation (between the recent segments) so as to maximize the join output rate, we use an equation similar to one in [17]. Let us start with a formula that captures the overall output rate of the join operation. Given the arrival rates of the streams λ_1 and λ_2, the smaller value of the selectivity factors of the recent segments σ, the output rate r_0 can be approximated as

$$r_0 = \sigma(\lambda_1|W_2^{rec}| + \lambda_2|W_1^{rec}|)$$
$$\text{Here, } |W_1^{rec}| + |W_2^{rec}| = M_{free}$$

The above equation can be rewritten as,

$$r_0 = \sigma(\lambda_1(M_{free} - |W_1^{rec}|) + \lambda_2|W_1^{rec}|)$$
$$= \sigma(\lambda_1 M_{free} + |W_1^{rec}|(\lambda_2 - \lambda_1))$$

Without loss of generality, let use assume that $\lambda_2 > \lambda_1$; hence, from the above equation, it is easy to deduce that the best strategy is to allocate most memory to the window corresponding to the lower input rate. To amortize the disk access cost over a large number of fresh tuples, while invoking disk probe, we set the minimum size of a recent segment as $M r_{rec}(0 < r_{rec} < 0.5)$, equivalent to N_{rec}^{min} blocks; Here, M is the total join memory. Here, N_{rec}^{min} is greater than N_{flush}.

To capture the instantaneous arrival rates of a stream, we use a metric termed as arrival frequency ($C_i; i = 1, 2$) that is decayed at the elapse of an interval or epoch (t_{epoch}. The decaying scheme [21] works as follows: within an epoch, the arrival frequency of a stream is increased upon the processing of a block from the stream; whereas, at the end of every epoch, the metric C_i is updated as αC_i. Stream i yields a block to the recent segment of stream $\bar{i}$, if the following conditions hold,

$$C_i \geq (1 + \rho)C_{\bar{i}}$$
$$|W_{\bar{i}}^{rec}| > B_{rec}^{min} \land \sum_{p=0}^{n_{part}-1}(flush_i^p + stale_i^p) \geq 0$$

In the above inequalities, ρ is a user defined parameter that controls the aggressiveness of eviction of a block within the recent segment. Within the recent segment, a victim partition is selected based on the productivity metric ($prod_i^x$) of the partitions: a partition with the lowest productivity metric is selected that yields a tailing, flushed block.

Procedure 4 presents the algorithm for evicting a block from a recent segment. The join algorithm invokes the procedure whenever a recent segment is full and new block is desired to store the incoming window-tuples. The procedure selects, based on the criteria presented in equation 5.5, a stream that should supply a block (line 1–4). The selected stream usually supplies an available flushed block; if the selected stream has no flushed blocks, the procedure either flushes the stale blocks onto the disk (creating

Procedure 4. EVICTRECENTBLOCK(int k)

Require: input: k = stream index ;
   return: a freed block b
   global variables: W_i; $i = 0, 1$

1: $i \leftarrow k$
2: **if** CANSUPPLY($\bar{k}$) **then**
3:    $i \leftarrow \bar{k}$
4: **end if**
5: **if** $flush_i = 0$ **then**
6:    **if** $W_i^{disk} = \phi$ **and** $free(W_i^{inv}) > 0$ **then**
7:       $p \leftarrow$ partition having the block with the lowest BN
8:       remove tailing block b in $W_i^{rec}[p]$
9:       copy b to $W_i^{inv}[p]$
10:       decrement $stale_i^p$
11:    **else**
12:       $n \leftarrow min(B_{rec}^{min}, \sum_{p=0}^{n_{part}-1} stale_i^p)$
13:       DISKDUMP(i,n)
14:    **end if**
15: **end if**
16: **if** $flush_i \neq 0$ **and** $free(W_i^{rec}) \leq 0$ **then**
17:    $p \leftarrow \arg\min_{x \mid stale_i^x > 0} (prod_i^x)$
18:    remove tailing block b in $W_i^{rec}[p]$
19:    decrement $flush_i^p$
20:    UPDATEFREQSEGMENT(W_i^{freq}, b)
21: **end if**
22: **return** b

Procedure 5. AH-EWJ

1: initialize variables $premature_i$, $freshN_i$, UI_i, $flush_i^p$, $stale_i^p$, BN_i {set to 0; i=0,1; $0 \leq p < n_{part}$}
2: **loop**
3:    retrieve a tuple s_i from input buffer of S_i in FIFO order
4:    $p \leftarrow \mathcal{H}(s_i)$
5:    $W_i^{rec}[p][H] \leftarrow \{W_i^{rec}[p][H] \cup s_i\}$
6:    **if** $W_i^{rec}[p][H]$ is full **then**
7:       compute $W_i^{rec}[p][H] \bowtie \{W_{\bar{i}}^{mem}[p]\}$
8:       $W_{rec(i)}[p][H].BN \leftarrow BN_i$
9:       increment BN_i
10:       increment $freshN_i$
11:       **if** W_i^{rec} is full **then**
12:          $W_i^{rec}[p] \leftarrow$ EVICTRECENT-BLOCK(i)
13:       **else**
14:          add a free block to the head of $W_i^{rec}[p]$
15:       **end if**
16:    **end if**
17:    **if** $freshN_i \geq rN_{rec}^{min}$ **then**
18:       $UI_i \leftarrow UI_i + freshN_i \times |b_i^m|$
19:       $freshN_i \leftarrow 0$ { reset $freshN_i$ }
20:       **if** $UI_i \geq Epoch_i$ **then**
21:          DISKPROBE(W_i^{rec}, $W_{\bar{i}}^{disk}$, W_i^{freq}, 1)
22:          $UI_i \leftarrow 0$ {reset the count}
23:       **else**
24:          DISKPROBE(W_i^{rec}, $W_{\bar{i}}^{disk}$, W_i^{freq}, 0)
25:       **end if**
26:       **for** $p \leftarrow 0$ to $n_{part} - 1$ **do**
27:          $stale_i^p \leftarrow |W_i^{rec}[p]| - flush_i^p$
28:       **end for**
29:    **end if**
30: **end loop**

a pool of flushed blocks) or copy a stale block to the invalidation segment. The former scenario arises if the respective disk segment is not empty or the invalidation segment has no free blocks; therefore, a pool of stale blocks are dumped onto the disk converting them to the flushed ones (line 12–13); whereas, in the latter situation, if the disk segment is empty, the stale blocks bypass the disk dump and are transferred to the invalidation segment, provided the invalidation segment has free space to accommodate the incoming stale blocks (line 7–10). The procedure evicts a flushed block (line 17–20) in two scenarios: (1) the pool of flushed blocks are initially empty, and the stale blocks, not being bypassed the disk segment as stated above, are flushed onto the disk; (2) the pool of flushed blocks are not empty initially (i.e., $\sum_{p=0}^{n_{part}-1} flush_i^p \neq 0$). Procedure DISKDUMP(i,n), in line 13, dumps n stale blocks of window W_i^{rec} onto the disk. The details of the procedure is described in 5.4. The procedure UPDATEFREQUENTSEG-MENT in line 20 attempts to replace a block within the frequent segment by the evicted flushed block depending on the productivity values of the blocks.

Table 1. Notations and the system parameters

Notations	Description
$S_i, S_{\bar{i}}$	stream i and opposite to i, respectively
$W_i, W_{\bar{i}}$	Window of stream S_i and $S_{\bar{i}}$, respectively
W_i^{disk}	Disk portion of window W_i
W_i^{rec}	recent segment of window W_i
W_i^{freq}	Frequent segment of window W_i
W_i^{inv}	invalidation segment of window W_i
W_i^{mem}	memory portion of window W_i, $(W_i^{rec} + W_i^{freq} + W_i^{inv})$
$W_i^{seg}[p]$	partition p of $seg \in \{rec, freq, inv\}$
$W_i^{rec}[p][H]$	block at the head of $W_i^{rec}[p]$
b_i^p	block p in window W_i
$b_i^p.minBN$	minimum block number from $W_{\bar{i}}^{disk}$ & $W_{\bar{i}}^{rec}$ that $b_i^p \in W_i^{rec}$ joins with
$b_i^p.prod$	Productivity of block b_i^p
ρ	parameter to evict a block within a frequent segment and a recent segment

5.6 Join Algorithm

Having described the details of the techniques behind the join algorithm, we now present the join algorithm AH-EWJ in Procedure 5. Within the join algorithm an infinite loop fetches, in each iteration, tuples from an input buffer and joins the fetched tuples with the opposite stream. At each step, the stream having a pending tuple/block with lower timestamp (i.e., the oldest one) is scheduled. A tuple fetched from the buffer is mapped to its partition and accumulated into the block at the head of the respective partition within the Recent segment. If the head-block of a partition p becomes full, it is joined with the memory portion of the partition ($W_i^{mem}[p]$). The filled block at the head of the partition is assigned a unique block number BN from an increasing sequence. The partition p is allocated a new head block (line 11–15). If no free block is available in the recent segment (W_i^{rec}), a victim window is selected; and a block from a selected partition of the the victim window segment is evicted (line 12).

The variable $freshN_i$ tracks the number of fresh blocks in W_i^{rec}. Whenever a recent segment is filled with at least N_{rec}^{min} fresh blocks, the disk probe phase in invoked (line 17). Blocks in Frequent segment W_i^{freq} are updated periodically at the end of an epoch $Epoch_i$; and this update phase is merged with a disk-probe. The epoch can be measured either by time or by the number of tuples processed. We use the later in specifying the epoch. A disk-probe with the update phase (in line 21) is invoked if the number of tuples processed (UI_i) since the last update exceeds the epoch length; otherwise, the *disk probing* omits the update phase setting the last parameter to 0 (line 24). At the end of the disk-probe, parameter $stale_i^p$ for each partition p is changed, converting the fresh blocks within the partition to stale ones (line 26–28).

6 Window Joins Using Variants of XJoin

In this section, we present a methodology to process window-based stream joins, which is hereafter termed as RPWJ (Rate-based Progressive Window Join), extending an existing algorithm RPJ [12](Rate-based Progressive Join)—a variant of XJoin [9]. Though RPJ is close candidate for processing sliding window joins, it suffers high disk I/O overhead. The RPWJ algorithm extends RPJ to reduce the overheads in processing sliding window joins incorporating disk storage. While joining two finite relations, RPWJ, like RPJ, maximizes the output rate by tuning the flushing policy and the reactive stage (i.e., md- or dd-task) based on the data distribution and tuple arrival patterns of the relations. However, as described in the subsequent part of this section, the temporal aspects of data within a sliding window invalidates the application of sort-merge join during a reactive stage; therefore, the estimation of the output rate of a md- or dd-task should be revisited. Moreover, we present a simplified policy to compute incrementally the output generation rate of both the md- and dd-tasks. This section starts with an overview of the RPWJ algorithm, presents the optimal flushing policy, and analyzes the estimation of the output rate of a reactive task.

6.1 Overview

In sliding window joins, as the window slides over time, tuples from should be invalidated periodically. This period depends on the sliding interval τ. In a continuously sliding window (i.e., τ is equal to the minimum unit or quanta of time), tuples from a window should be invalidated whenever a new tuple arrives in the opposite stream. Such an eager invalidation is infeasible while processing sliding window joins using a disk archive, as it would require a disk access for each incoming tuple. In such a scenario, lazy invalidation should be used. However, a tuple can be invalidated only if it is joined with all the tuples from the opposite window. Thus, the reactive stages (both md- and dd-stage) must be carried out immediately before an invalidation operation within a partition. Carrying out the reactive stages for all the partitions is a time consuming process. Thus the interval between two successive invalidations should be high. Note that even with a low stream arrival rate, such a lazy invalidation might result in a significant number of output tuples produced outside the time limit set by the window. [1]

---

[1] Such tuples are reintegrated with the join results by a re-integration module outside the join module.

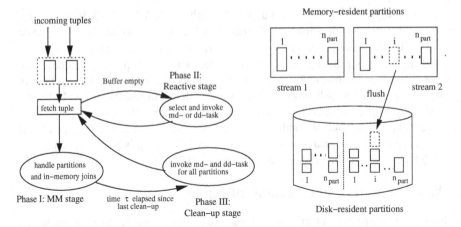

Fig. 5. Architecture of the RPWJ algorithm

Fig. 6. Memory stage of the join algorithm

The RPWJ should be applied to window joins with a large τ. Having the shortcoming of the SPWJ, we now describe the algorithm. SPWJ is carried out in three stages performed in an interleaved fashion (Figure 5).

The first stage joins the incoming tuples with the memory-resident data of the opposite stream. SPWJ organizes the tuples within window W_i of stream S_i into partitions (Figure 6). Each partition j ($1 \leq j \leq n_{part}$, n_{part} being the total hash partitions) consists of two parts: a memory-resident part $W_i^{mem}[j]$, that stores the recent data within the partition, and a disk-resident part $W_i^{mem}[j]$, that stores data flushed to the disk due to memory limitations. When input tuples arrive from a source, the tuples are mapped to the respective partitions and joined with the memory-resident part of the partitions from the opposite window. If there is enough memory to store the tuples, then the tuples are stored in memory. Otherwise, one or more memory resident partitions are flushed onto the disk to make rooms for the incoming tuples. The first stage continues as long as there exist some tuples in the buffer to process; otherwise, the algorithm continues to the second phase.

The second phase chooses a reactive task from a pool of at most $3n_{part}$ possible tasks: there are at most $2n_{part}$ md-tasks ($W_i^{mem}[j] \bowtie W_{\bar{i}}^{disk}[j]$ for all $1 \leq j \leq n_{part}$ and $i = 1, 2$) and n_{part} dd-tasks ($W_1^{disk}[j] \bowtie W_2^{disk}[j]$ for all $1 \leq j \leq n_{part}$). The algorithm chooses a candidate task using a metric *Expected Output Rate*, which is defined as Er/Et, where Er is the expected number of output tuples generated by the task, and Et is the task execution time.

The third stage starts periodically at an interval of τ'. This stage invokes all $3n_{part}$ possible reactive tasks as stated above.

6.2 Join-Processing Granularity

In all previous approaches, the incoming tuples are usually processed a tuple at a time. However, we observe that, when the buffer contains enough pending tuples, the join

operation can be made more efficient by grouping the tuples in a block and processing the join operation at the granularity of the block. Such a block level join operation
reduces the number of scans of the respective memory partition. In addition to the overhead of partition scan, the block-level processing reduces the amount of meta-data to
be kept within a block. We switch between block-level and tuple-level processing depending on the buffer occupancy: whenever the buffer occupancy is above H_{mark}, the
algorithm processes at block-level; and while the buffer occupancy is below L_{mark}, it
switches to tuple-level.

Each block contains a binary field *gLevel* which is set to 0 if the tuples within the
block are processed at the tuple-level, and set to 1 otherwise. Duplicate elimination
within output results, as described in the subsequent part of the section, requires each
tuple or block to be associated with its arrival and departure (flush) timestamps. The departure timestamp (DTS) can be maintained at a block level irrespective if its processing
granularity. However, in case of the tuple-level processing, the arrival timestamp(ATS)
should be maintained at the tuple level. If tuples within a block b are processed at a
granularity of a tuple, the block contains an array of timestamps. The i-th element of
the array (ATS_i) contains the arrival timestamp of the i-th tuple within the block. In
the subsequent part of the section, we consider only the block level processing; and the
case for a block b with *gLevel* flag set to 0 can easily be extended with an extra level
of iteration over each tuple t_i (i.e., i-th tuple within the block) within the block and
checking against the relevant interval $[b.ATS_i, b.DTS]$.

6.3 Flushing Policy

The flushing policy, upon memory overflow, moves to the disk some memory resident
tuples that are expected to produce the least number of output tuples until the next
memory overflow. The victim partitions to be flushed are selected based on the arrival
frequencies $(C_i[p])$ of the partitions. The arrival frequency $(C_i[p])$, which is maintained
and decayed using a scheme similar to that in [21], indicates the volume of recent
arrivals of tuples within partition p of stream i.

The flushing algorithm sorts the arrival frequencies of the partitions in descending
order. The algorithm selects the arrival frequency $C_{i'}[p']$ with the minimum value, and
flushes tuples from the p'-th partition of the opposite window onto the disk. Starting
from the minimum arrival frequency, the algorithm continues to select and flush the
successive partitions as long as the total volume of the tuples flushed onto the disk is
less than N_{flush}. We flush the tuples at a partition level (i.e., don't flush a fraction of a
partition) to render the task of the incremental maintenance of the Er-values simpler.

6.4 Duplicate Elimination in Reactive Stages

The reactive stage starts when there is no input tuple to process, and selects one of the
md- or dd-tasks with the maximum output rate from a pool of $2n_{part}$ md-tasks and
n_{part} dd-task.

The md-stage joins a memory partition $W_i^{mem}[p]$ with the disk partition of the opposite stream window $(W_{\bar{i}}^{disk}[p])$. Within this task, blocks from the disk partition are read
and probed into the respective memory partition of the opposite window. The md-stage

may generate redundant output tuples. There are only two ways in which an output tuple might be redundant: (1) the output tuple was generated in the mm-stage, or (2) the output tuple was generated in a previous md-task that involved $W_i^{mem}[p]$ and $W_{\bar{i}}^{disk}[p]$. Each memory partition $W_i^{mem}[p]$ maintains a list $T_{i,p}^{md}$ that stores the timestamps of the previous md-task involving the memory partition. As the window slides over time, the list is pruned to remove the tailing timestamps that fall outside the time window. So, after an invalidation, if the memory partition $W_i^{mem}[p]$ was involved in c md-tasks, the list $T_{i,p}^{md}$ contains c entries, the last entry $T_{i,p}^{md}[c]$ indicating the most recent md-task involving the memory partition. Thus output tuples resulting from a block $b_d \in W_{\bar{i}}^{disk}[p]$ and a block $b_m \in W_i^{mem}[p]$ are propagated (i.e., not duplicate) if none of the following two conditions are satisfied:

1. $[b_d.ATS, b_d.DTS]$ intersects $[b_m.ATS, b_m.DTS]$, which means that b_m and b_d found each other in memory, and are already joined.
2. $T_{i,p}^{md}[c] > b_m.ATS$ and $b_d.DTS < T_{i,p}^{md}[c]$, which means that b_m was joined with b_d is a previous md-stage.

It should be noted that the values $T_{i,p}^{md}[1], T_{i,p}^{md}[2], \ldots, T_{i,p}^{md}[c-1]$ are not used in the duplicate elimination during a md-stage as stated above; however, these values are used during a dd-stage as stated below, and, therefore, should not be discarded.

The dd-task joins the two disk portions, from both the stream windows, of a partition p. The two disk partitions ($W_i^{disk}[p]$ and $W_{\bar{i}}^{disk}[p]$) are joined using the Blocked NLJ (Nested Loop Join) algorithm. The order of the tuples or blocks within a partition can't be altered. So, the Progressive Sort Merge Join (PSMJ) Algorithm, used while joining finite streams or relations, can no longer be applied in the scenario of a window join. We assume that the disk partition $W_i^{disk}[p]$ is used as the outer partition and the partition $W_{\bar{i}}^{disk}[p]$ as the inner one in the Block-NLJ algorithm. To reduce the number of disk I/O operations during the join, we read the disk blocks as clusters and store these blocks in a reserved memory space. While joining a block $b_i \in W_i^{disk}[p]$ and a block $b_{\bar{i}} \in W_{\bar{i}}^{disk}[p]$, there might arise duplicate tuples in the following three possible ways. In the first case, the two blocks might have been joined in memory, during the mm-stage. This can be detected by simply checking, similar to a md-stage, the memory-alive intervals of the two blocks.

The second scenario is that the blocks were joined in a previous md-stage, which could be a join between $W_i^{mem}[p]$ and $W_{\bar{i}}^{disk}[p]$, or a join between $W_i^{disk}[p]$ and $W_{\bar{i}}^{mem}[p]$. Due to the symmetry, we consider the only former one. In this case, there must be a md-task when the block b_i was in memory (i.e., within the interval $[b_i.ATS, b_i.DTS]$), but $b_{\bar{i}}$ was flushed to the disk (i.e., the md-task happened after $b_{\bar{i}}.DTS$). If there are a total of c_i md-tasks since the last invalidation of the stream window, then their timestamps are given by $T_i^{md}[1], T_i^{md}[2], \ldots, T_i^{md}[c_i]$, as stated earlier in the section. If any of the timestamps lies within the interval $[b_i.ATS, b_i.DTS]$ and is greater that $b_{\bar{i}}.DTS$, then the blocks b_i and $b_{\bar{i}}$ were already joined.

As the third case, the two blocks b_i and $b_{\bar{i}}$ might have been produced during a preceding dd-task between $W_i^{disk}[p]$ and $W_{\bar{i}}^{disk}[p]$. To track whether two blocks are already joined in a dd-task, we keep a timestamp $lastDD_i^p$, which tracks the timestamp of the most recent dd-task, for each disk partition $W_i^{disk}[p]$ of stream i. If $b_i.DTS$ is greater

$lastDD_i^p$ and $b_{\bar{i}}.DTS$ is greater $lastDD_{\bar{i}}^p$, then these two blocks are already joined during the last dd-stage invoking the disk parts $W_i^{disk}[p]$ and $W_{\bar{i}}^{disk}[p]$.

6.5   Task Selection

As mentioned earlier in this section, the output rate of a md- or dd-task is given by Er/Et, where Er is the expected number of new results to produced by a task, and Et is the task execution time. Predicting the output rate requires us to predict both Er and Et for all partitions of the stream windows. We denote the Er-value of an md-task between partitions $W_i^{mem}[p]$ and $W_{\bar{i}}^{disk}[p]$ as $Er_{i,p}^{mem}$, and that of a dd-task between $W_i^{disk}[p]$ and $W_{\bar{i}}^{disk}[p]$ as Er_p^{disk}. We start with the analysis of Et.

Analysis of Et.   For an md-task joining $W_i^{mem}[p]$ and $W_{\bar{i}}^{disk}[p]$, Et it is approximated by the time required to scan the disk partition $W_{\bar{i}}^{disk}[p]$—the time to access the in-memory partition $W_i^{mem}[p]$ is negligible compared to the access time of the disk partition. Blocks flushed to a disk partition, during optimal flushing, are stored sequentially on the disk; such a pool of blocks flushed together constitute a run as termed in [12, 14]. If blocks are flushed in $W_{\bar{i}}^{disk}[p]$ $x_{\bar{i}}$ times during the span of the stream window, it requires at least $x_{\bar{i}}$ random accesses. Suppose the size of the reserved memory space for disk I/O is M_{read}. Then, there will be at least $\frac{|W_{\bar{i}}^{disk}[p]|}{M_{read}}$ disk seek and latencies (disk accesses) even if the blocks of the disk partition are stored sequentially (i.e., only one run). Thus there will be at most $\left(x_{\bar{i}} + \frac{|W_{\bar{i}}^{disk}[p]|}{M_{read}} \right)$ disk accesses. Hence,

$$Et_{md} = \left(x_{\bar{i}} + \frac{|W_{\bar{i}}^{disk}[p]|}{M_{read}} \right).t_A + C_R * W_{\bar{i}}^{disk}[p] \tag{1}$$

Here t_A is the disk access time, C_R is the disk transfer rate expressed in sec/byte.

For a dd-task joining the partitions $W_i^{disk}[p]$ and $W_{\bar{i}}^{disk}[p]$, we need to fully scan the inner partition (say $W_{\bar{i}}^{disk}[p]$) for each iteration of the outer partition (i.e., $W_i^{disk}[p]$). As stated earlier, we share a reserved memory space of size M_{read} to read blocks from both the disk partitions. As analyzed in [24], for a block *nested loop join* involving two disk relations, the total disk I/O times is minimized when blocks read from disk at a time (i.e., *cluster size* [24]) is $M_{read}/2$ (i.e., the available memory is evenly divided between the relations). Hence, while scanning each of the disk partitions, we read $M_{read}/2$ blocks at a time. Thus with the cluster size $M_{read}/2$, the total time to scan a partition $W_i^{disk}[p]$ is given by

$$Et_{i,p} = \left(x_i + \frac{2|W_i^{disk}[p]|}{M_{read}} \right).t_A + C_R * W_i^{disk}[p] \tag{2}$$

Also, with the cluster size $M_{read}/2$, the inner partition $W_{\bar{i}}^{disk}[p]$ should be scanned $2|W_i^{disk}[p]|/M_{read}$ times. Therefore, Et for the dd-task can be written as,

$$Et_{dd} = Et_{i,p} + \frac{2|W_i^{disk}[p]|}{M_{read}} Et_{\bar{i},p} \tag{3}$$

Here, $Et_{i,p}$ and $Et_{\bar{i},p}$ are derived from equation 2.

Analysis of Er. As new tuples arrives, the expected number of output results from a reactive task changes over time. We provide a technique to estimate the expected join results in a md or dd-task. The Er-values for the partitions changes whenever a new tuple or block arrives in memory or a memory block is flushed onto the disk.

When a new tuple (in tuple-level processing) or block (in block-level processing) arrives in the partition p of stream i, the arriving tuple tuple or block will be joined, during a subsequent md-task, with all the blocks in the disk portion of partition p of the opposite window $W_{\bar{i}}^{disk}[p]$. Thus, an arriving block b_m (considering the block-level processing) should generate $\sigma.|b_m|.|W_{\bar{i}}^{disk}[p]|$ output tuples, σ is the join selectivity within partition p. As stream tuples are partitioned using a hash function, we assume, like [12], the same join selectivity σ over all partitions. Upon arrival of a block b_m, the expected output tuples ($Er_{i,p}^{md}$) for an md-task involving $W_i^{mem}[p]$ and $W_{\bar{i}}^{disk}[p]$ is increased by $\sigma.|b_m|.|W_{\bar{i}}^{disk}[p]|$. Whenever an md-task $W_i^{mem}[p] \bowtie W_{\bar{i}}^{disk}[p]$ is invoked, the value of $Er_{i,p}^{md}$ is reset to zero. Also, whenever a memory partition p of stream i ($W_i^{mem}[p]$) is flushed onto the disk, the respective Er-value (i.e., $Er_{i,p}^{mem}$) is reset to zero. However, note that the output tuples to be generated in a md-task $W_i^{mem}[p] \bowtie W_{\bar{i}}^{disk}[p]$ will now be generated within a subsequent dd-task involving partition p. Thus, before resetting to zero, the value $Er_{i,p}^{md}$ is added to the Er_p^{dd}, which is the expected output tuples from a dd-task $W_i^{disk}[p] \bowtie W_{\bar{i}}^{disk}[p]$.

As disk flushing is carried out at the granularity of partitions, the task of maintaining the Er-values of the dd-tasks becomes simple. Such partition-level flushing does not affect the join performance, as the optimal performance is not restricted to a fixed value of N_{flush} but a wide range of values for N_{flush}. Note that the expected output rates of various tasks are compared among themselves while choosing an optimal reactive task, and their absolute values are not necessary. So, we can omit the join selectivity σ from the above calculation and use the values of $Er_{i,p}^{md}$ or Er_p^{dd}—which now provide the size of the Cartesian products—as an indication of the relative values of their output size.

7 Experiments

This section describes our methodologies for evaluating the performance of the AH-EWJ algorithm and presents experimental results demonstrating the effectiveness of the proposed algorithm. We begin with an overview of the experimental setup.

7.1 Simulation Environment

We evaluated the performance of the prototype implementation using synthetic traces. All the experiments are performed on an Intel 3.4 GHz machine with 1GB of memory. We implemented the prototype in Java. As a buffer, we allocated 2MB of memory and divided this memory between the streams. We incorporated the disk as a secondary storage while buffering the arriving tuples [25]. Here, it should be noted that we have not measured the processing delay for the tuples in buffers. The main focus of our experimentation is to observe the effectiveness or the efficiency of the join processing algorithm. The number of the buffered tuples indicates when the system is overloaded or saturated; in such a scenario, the number of buffered tuples increases with time. Thus the size of the buffer is virtually unbounded and plays no role in the experiments.

Traces. We generate the time-varying data streams using two existing algorithms: PQRS [26] and *b-model* [27]. PQRS algorithm models the spatiotemporal correlation in accessing disk blocks whereas the b-model captures the burstiness in time sequence data. In the experiments, we generate data streams for a certain duration T. We observe that the time to generate a large volume of data tuples within T while using the PQRS algorithm is prohibitively large as it requires us to sort the tuples based on their times-tamps. We divide the duration T into 2^n mini-intervals, measure the total volume of tuples within each mini-interval using *b-model*. We choose the aggregation level (n) in *b-model* so as to bound the maximum tuple volume within a mini-interval to a given value: this upper limit varies with the parameter b in *b-model* (e.g., for b=0.6 and T = 1.6 hours, we set the upper limit to 70000 tuples). Having determined the total volume of tuples within a mini-interval, these tuples are generated using the PQRS algorithm. In our settings, the average arrival rate (for a stream) denotes the long term average within T, and the instantaneous arrival rates depends on the parameters capturing burstiness both in PQRS and *b-model*.

System Parameters, Metrics and Default Values. To model the access time for the disk segment, we divide a disk segment (W_i^{disk}) into n_i basic windows $B_{ij}^{win}(j = 1 \ldots n_i)$ [28]. We assume that disk blocks within a basic window are physically con-tiguous, and the delay in accessing the blocks in a basic window is estimated only by the disk bandwidth(i.e., no seek or latency). However, accessing blocks beyond the boundary of a basic window imparts an extra delay equivalent to the seek time and rotational latency (i.e., an access time). We set the size of the basic window to 1MB. We allocate 70% of memory to the recent segments. The remaining memory is equally distributed among the invalidation and frequent segments. Minimum size of a recent segment (B_{rec}^{min}) is set to a fraction 0.3 of the total memory reserved for the recent seg-ments. We set the minimum delay between successive reactive stages (both for RPWJ and AH-EWJ) as 15sec. Note that, for RPWJ, the reactive stages occur at a partition level, whereas AH-EWJ carries out reactive stage (i.e., disk probe while CPU is idle) at a stream level. As a base device, we consider IBM 9LZX and use its parameters in measuring the access time.

In our experiments, we fix the memory page and also the disk block size to 1KB. Each record has a length of 64 bytes. The domain of the join attribute A is taken as integers within the range $[0 \ldots 10 \times 10^6]$. We allocate 2MB of memory to buffer the stream tuples generated up to a particular instant. We divide the memory between the buffers in the two streams. When a buffer overflows, we dump the incoming tuples onto the disk. Hence, the total number of the buffered tuples indicates the degree of overload of the system. In normal cases, the buffer size or the volume of pending tuples increases within a bursty interval and decreases while the load is not very high. But, when the system in permanently overloaded, the volume of buffered tuples increases with time saturating the whole system.

In addition to the total pending or buffered tuples, we measure *average production delay*, total CPU time, total disk time and maximum total window size. We measure the delay in producing an output tuple as the interval elapsed since the arrival of the joining tuple with more recent timestamp. For example, if tuples s_1 and s_2 are the joining tuples of the output tuple (s_1, s_2), where $s_1.t > s_2.t$ (s_1 being the more recent) and current

Table 2. Default values used in experiments

Parameter	Defaults	Comment
$W_i(i = 1, 2)$	10	Window length (min)
δ	1	slide interval for a Window (min)
λ	1200	Avg. arrival rate(tuples/sec)
α	0.4	Decay parameter for the productivity value
b	0.6	burstiness in traces (captured by *b-model*)
ρ	0.4	block eviction threshold
M	20	join memory (MB)
N_{flush}	$0.4M$	flush size for RPWJ (MB)
n_{part}	60	hash partitions or buckets

time is T_{clock}, then the delay in producing the output tuple is $(T_{clock} - s_1.t)$. This metric (i.e., average production delay) indicates how quick an output tuple is generated; hence, can be considered as an indication of instantaneous output generation rate. We also measure the percentage of *delayed tuples* that indicates the fraction of tuples not processed within the time limit of the stream window. It happens due to the delay in bringing the blocks from a disk segment to the invalidation segment.

Unless otherwise stated, the default values used in the experiments are as given in Table 2.

7.2 Experimental Results

In this section, we present a series of experimental results for assessing the effectiveness of the proposed join algorithm. As a baseline algorithm, we use the RPWJ (Rate-based Progressive Window Join) algorithm as presented in Section 6. The extension is imperative, as RPJ joins only the finite streams and suffers from large I/O overheads while processing sliding window joins. For each set of the experimentations, we run the simulator for 1.6 simulation hours. We start to gather performance data after an startup interval of 12 minutes is elapsed.

Varying Arrival Rates. We, now, present the experimental results with varying stream arrival rates. Figure 7 and Figure 8 show the total buffered tuples (from both the joining streams) at different time-points for the algorithm AH-EWJ and RPWJ, respectively. As shown in the figures, the number of the buffered tuples increases with the increase in arrival rates. In case of the RPWJ, the system is overloaded for a per-stream arrival rate of 1400 tuples/sec.

With the increase in the arrival rates, more and more tuples can not be processed within the time limit of the stream window; thus, the percent of delayed tuples increases with the increase in the arrival rates as shown in Figure 9. The tuples that cannot be processed within the time limit of the stream window is denoted as *delayed tuples*. These delayed tuples get expired at a time later than the usual after they get joined with the respective window. For RPWJ, the percentage of the delayed tuples increases

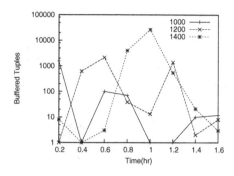

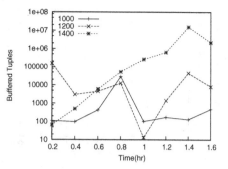

Fig. 7. Buffer size (for different stream arrival rates) at different time points (AH-EWJ)

Fig. 8. Buffer size (for different stream arrival rates) at different time points (RPWJ)

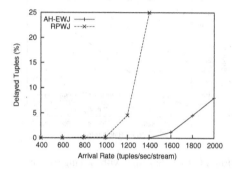

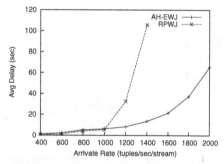

Fig. 9. Percentage of delayed tuples with varying stream arrival rates

Fig. 10. Average delay with varying stream arrival rates

sharply with the increase in arrival rates beyond 1000 tuples/sec; however, in case of the AH-EWJ, this percent of delayed tuples remains low even for an arrival rate of 1800 tuples/sec/stream (i.e., 3600 tuples/sec in the system). Here, it should be noted that *the load applied to the join module is a function of both the window size and the arrival rate; therefore, mere the arrival rate by itself provide little indication regarding the system load.*

Figure 10 shows, for both the algorithms, the average delay in producing output tuples with the increase in arrival rates. Figure 11 shows the maximum window size during the system activity. Though the allocated memory per stream window is 20MB, spilling the extra blocks onto the disk does not impart significant increase in average output delay of the AH-EWJ even for arrival rates up to 1600 tuples/sec/stream, at a point where maximum total window size is around 190MB (i.e., 9 times the join memory size). Techniques based on load shedding would have discarded the extra blocks beyond 20MB losing a significant amount of output tuples that would never have been generated. The RPWJ algorithm becomes saturated for an arrival rate 1400 tuples/sec/stream at a point where the average delay attains a very high value. Hence,

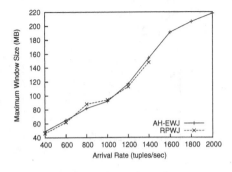

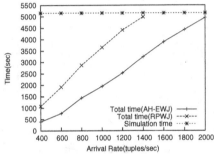

Fig. 11. Maximum window size (MB) with varying stream arrival rates

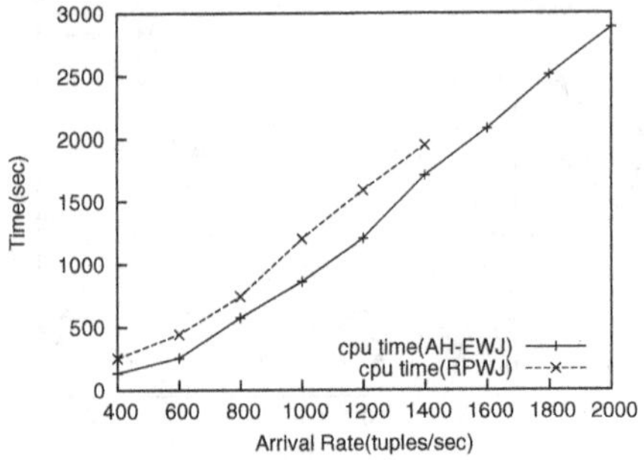

Fig. 12. Total cpu time with varying stream arrival rates

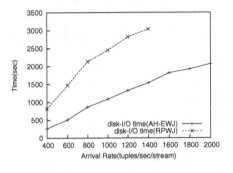

Fig. 13. Total disk time with varying stream arrival rates

Fig. 14. Total disk time with varying stream arrival rates

for a large window, the proposed technique attains a low average delay of output generation; at the same time, the percentage of the tuples missing the time frame set by the window is very low (0.05% for an arrival rate 1400 tuples/sec/stream). This demonstrates the effectiveness of the AH-EWJ algorithm.

Figure 12 shows the CPU time for varying arrival rates. The CPU time for the RPWJ is higher while compared to the AH-EWJ. This difference in CPU times can be attributed to the frequent invocation of the reactive stages that don't consider the amortization of a partition-scan over a large number of pending tuples. Figure 13 shows the disk I/O time with varying arrival rates. The disk overhead of the RPWJ is very high while compared to that of AH-EWJ algorithm. In RPWJ the reactive stages are invoked for each individual partition, an approach that results in expensive disk I/Os for a small number of input tuples. Figure 14 shows the total system time (disk I/O and CPU time) with varying per stream arrival rates. We notice that with the RPWJ algorithm the system is saturated at the rate of 1400 tuples/sec/stream. On the other hand, with the same arrival rate, the AH-EWJ algorithms achieves around 40% reduction in total execution time.

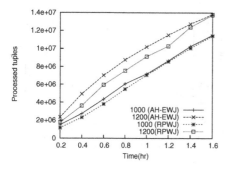

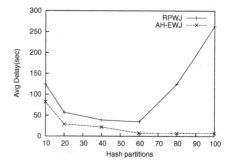

Fig. 15. Total processed tuples at various time points

Fig. 16. Average delay in generating output tuples with varying sizes of hash partitions

Figure 15 shows the processed tuples at difference time points. For a fixed arrival rate, the total tuples processed by the two algorithms are same. The pattern of tuple arrivals over time is almost similar for both of the algorithms. However, for same arrival rates, the processed tuples of the two algorithms at various time points might deviate due to the variations in the location of bursts in time.

Varying Hash Partitions. Figure 16 shows the effect of increasing hash partitions on the performance of the two algorithms. For RPWJ, increasing the hash partitions beyond a certain values (i.e., 60) results in increased average delay. For RPWJ, as the number of hash partitions increases, the disk overhead also increases. This disk overhead occurs while flushing the tuples, expiring the tuples, and probing disk partitions. As stated earlier, algorithms based on sorting the tuples and merging the sorted runs can't be applied in processing sliding window joins. So, increasing the number of hash partitions can be approach to lower the processing overhead. But, the performance of the RPWJ algorithm deteriorates with the increase in the hash partitions; this phenomenon renders the RPWJ unfitting for the sliding window joins. On the other hand, AH-EWJ attains lower average delay with an increase in the hash partitions.

Varying Bias. Figure 17 shows the average delay in generating output tuples with varying bias or burstiness of the data stream. As shown in the figure, with the increase in the bias, AH-EWJ performs significantly better than the RPWJ. Figure 18 shows the maximum window sizes with varying bias or skewness of stream arrivals. With the increase in bias, the maximum window size also increases. Figure 19, and Figure 20 shows the CPU time, disk I/O time, and total System time (disk I/O and CPU time), respectively, with varying skewness in the stream arrivals. As shows in the figures, in case of the RPWJ, both the CPU time and disk I/O time decrease with the increase in the skewness. With the increase in the bias, during a bursty interval there occurs, if any, a very few reactive stages. Thus the scan of a stream window partition (while joining the incoming tuples) and disk accesses (to retrieve disk resident stream tuples) are amortized over a larger number of stream tuples. This phenomenon reduces both the processing overheads and the disk I/O time. It is interesting to note that although

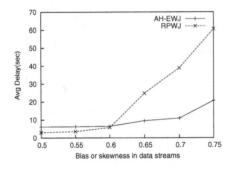

Fig. 17. Average delay in generating output tuples with varying bias

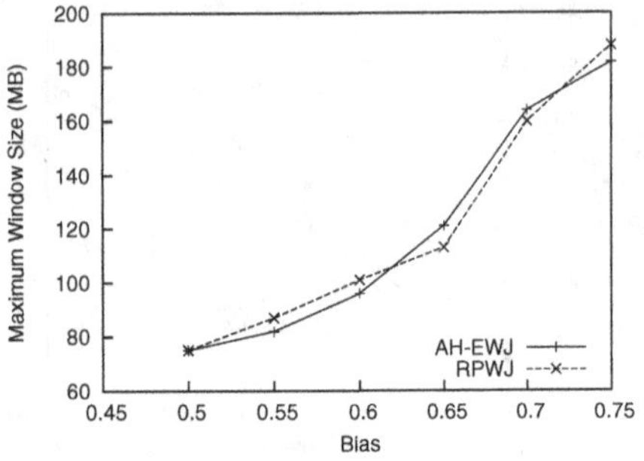

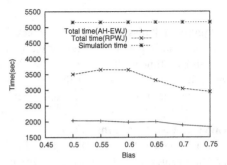

Fig. 18. Maximum window size with varying bias

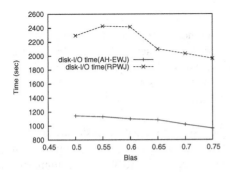

Fig. 19. Disk-I/O time with varying bias

Fig. 20. Total execution time with varying bias

the maximum window size increases with an increase of the bias, the total processing time decreases due to the amortization of the both the disk I/O and the window scan over a larger number stream tuples. Such an amortization outperforms the processing overhead due to the increase in the window size.

Varying Memory Size. Figure 21 shows, for AH-EWJ algorithm, the average delay in output tuples with varying memory size. With the increase in the allocation of memory to process the join operator, the average delay decreases. Also, increasing memory allocation to the join operator decreases both the CPU time and disk I/O time as shown in Figure 22. With the increase in join memory, the number of blocks spilled onto the disk also decreases, which decreases the disk I/O time during the disk probes. With the decrease in the volume of tuples participating the disk probes, the total processing time also decreases: scanning a window partition during a disk probe requires the partition to be loaded into the cache, increasing the processing overhead.

7.3 Comparison with Relevant Research

In our research, we consider the problem of producing exact results for stream joins given a limited memory. Existing approaches shed load during peak workload by

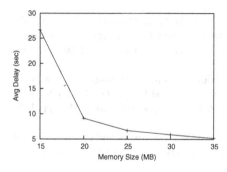

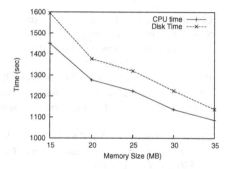

Fig. 21. Average delay in generating output tuples with varying memory

Fig. 22. Total CPU and disk time with varying memory

dropping tuples(e.g., [6, 5, 7], and/or processing (e.g., [28]). These approaches fail to produce exact results within an environment where data arrival rate is bursty in nature and/or the length of the stream windows exceeds the available memory. Our scheme handles the burstiness in stream arrivals by spilling data on the disk upon buffer overflow; and it copes with large window by spilling window data on the disk.

The experimental results show that the generation of exact results with bounded average production delay is feasible. We observe, for a window length of 10 min, that the fraction of delayed tuples processed to refine the output results remains very low ($\approx 0.1\%$) even for an arrival rate of 1600 tuples/sec/stream in the system.

8 Conclusion

In this paper, we address the issue of processing exact, sliding window join between data streams. We provide a framework for processing the exact results of an stream join, and propose an algorithm to process the join in a memory limited environment having burstiness in stream arrivals. Storing the stream windows entirely in memory is infeasible in such an environment. To the best of our knowledge, AH-EWJ is the first algorithm proposed to deal with such a scenario. Like any online processing, maximizing output rate is a major design issue in processing exact join. Hence, we propose a generalized framework to keep highly productive blocks in memory and to maintain the blocks in memory during systems activity, forgoing any specific model of stream arrivals (e.g., age-based or frequency-based model [5]). Moreover, the AH-EWJ eliminates the expensive disk-to-disk phase, and at the same time, amortize a disk scan over a large number of input tuples, rendering the algorithm IO-efficient. The algorithm minimizes disk dump by adapting the sizes of both the windows and the partitions. The experimental results demonstrate the effectiveness of the algorithm.

References

1. Carney, D., Çetintemel, U., Cherniack, M., Convey, C., Lee, S., Seidman, G., Stonebraker, M., Tatbul, N., Zdonik, S.B.: Monitoring streams – a new class of data management applications. In: Proc. Int. Conf. on Very Large Databases, VLDB, Hong Kong, China, pp. 215–226 (August 2002)

2. Chandrasekaran, S., Cooper, O., Deshpande, A., Franklin, M.J., Hellerstein, J.M., Hong, W., Krishnamurthy, S., Madden, S.R., Raman, V., Reiss, F., Shah, M.A.: TelegraphCQ: Continuous dataflow processing for an uncertain world. In: Proc. Conf. on Innovative Data Systems Research, CIDR (January 2003)

3. Babcock, B., Babu, S., Datar, M., Motwani, R., Widom, J.: Processing sliding window multi-joins in continuous queries over data streams. In: Proc. ACM SIGMOD-SIGACT-SIGART Symposium on Principles of Database Systems, PODS, Madison, Wisconsin, USA, pp. 1–16 (June 2002)

4. Gedik, B., Wu, K.-L., Yu, P.S., Liu, L.: A load shedding framework and optimizations for m-way windowed stream joins. In: Proc. Int. Conf. on Data Engineering, Istanbul, Turkey, pp. 536–545 (April 2007)

5. Srivastava, U., Widom, J.: Memory-limited execution of windowed stream joins. In: Proc. Int. Conf. on Very Large Databases, VLDB, Toronto, Canada, pp. 324–335 (September 2004)

6. Das, A., Gehrke, J., Riedewald, M.: Approximate join processing over data streams. In: Proc. ACM SIGMOD Int. Conf. on Management of Data, San Diego, USA, pp. 40–51 (June 2003)

7. Tatbul, N., Çetintemel, U., Zdonik, S.B., Cherniack, M., Stonebraker, M.: Load shedding in a data stream manager. In: Proc. Int. Conf. on Very Large Databases, VLDB, Berlin, Germany, pp. 309–320 (September 2003)

8. Liu, B., Zhu, Y., Rundensteiner, E.A.: Run-time operator state spilling for memory intensive long-running queries. In: Proc. ACM SIGMOD Int. Conf. on Management of Data, Chicago, Illinois, USA, pp. 347–358 (June 2006)

9. Urhan, T., Franklin, M.J.: XJoin: A reactively-scheduled pipelined join operator. IEEE Data Engineering Bulletin 23(2), 7–18 (2000)

10. Mokbel, M., Liu, M., Aref, W.: Hash-merge-join: A non-blocking join algorithm for producing fast and early join results. In: Proc. Int. Conf. on Data Engineering, pp. 251–263 (2004)

11. Viglas, S.D., Naughton, J.F., Burger, J.: Maximizing the output rate of multi-way join queries over streaming information sources. In: Proc. Int. Conf. on Very Large Databases, VLDB, Berlin, Germany, pp. 285–296 (September 2003)

12. Tao, Y., Yiu, M.L., Papadias, D., Hadjieleftheriou, M., Mamoulis, N.: RPJ: Producing fast join results on streams through rate-based optimization. In: Proc. ACM SIGMOD Int. Conf. on Management of Data, Baltimore, Maryland, USA, pp. 371–382 (June 2005)

13. Wilschut, A.N., Apers, P.M.G.: Dataflow query execution in a parallel main-memory environment. In: Proc. Int. Conf. on Parallel and Distributed Information Systems, PDIS, Miami, Florida, USA, pp. 68–77 (December 1991)

14. Dittrich, J.-P., Seeger, B., Taylor, D.S., Widmayer, P.: Progressive merge join: A generic and non-blocking sort-based join algorithm. In: Proc. Int. Conf. on Very Large Databases, VLDB, Hong kong, China, pp. 299–310 (August 2002)

15. Levandoski, J., Khalefa, M.E., Mokbel, M.F.: Permjoin: An efficient algorithm for producing early results in multi-join query plans. In: Proc. Int. Conf. on Data Engineering, Cancun, Mexico, pp. 1433–1435 (2008)

16. Double Index NEsted-Loop Reactive Join for Result Rate Optimization (2009)

17. Kang, J., Naughton, J.F., Viglas, S.: Evaluating window joins over unbounded streams. In: Proc. Int. Conf. on Data Engineering, Bangalore, India, pp. 341–352 (March 2003)

18. Ojewole, A., Zhu, Q., Chi Hou, W.: Window join approximation over data streams with important semantics. In: Proc. Int. Conf. on Information and Knowledge Management, CIKM, Virginia, USA, pp. 112–121 (November 2006)

19. Golab, L., Ozsu, T.: Processing sliding window multi-joins in continuous queries over data streams. In: Proc. Int. Conf. on Very Large Databases, VLDB, Berlin, Germany, pp. 500–511 (September 2003)

20. Teubner, J., Mueller, R.: How soccer players would do stream joins. In: Proc. ACM SIGMOD Int. Conf. on Management of Data, pp. 625–636 (2011)
21. Chakraborty, A., Singh, A.: A partition-based approach to support streaming updates over persistent data in an active data warehouse. In: Proc. IEEE Int. Symp. on Parallel and Distributed Processing, IPDPS, Rome, Italy, pp. 1–11 (May 2009)
22. Chakraborty, A., Singh, A.: A Disk-Based, Adaptive Approach to Memory-Limited Computation of Windowed Stream Joins. In: Bringas, P.G., Hameurlain, A., Quirchmayr, G. (eds.) DEXA 2010, Part I. LNCS, vol. 6261, pp. 251–260. Springer, Heidelberg (2010)
23. Babu, S., Munagala, K., Widom, J., Motwani, R.: Adaptive caching for continuous queries. In: Proc. Int. Conf. on Data Engineering, Tokyo, Japan, pp. 118–129 (April 2005)
24. Graefe, G.: Query evaluation techniques for large databases. ACM Computing Surveys 25(2), 73–169 (1993)
25. Motwani, R., Thomas, D.: Caching queues in memory buffers. In: Proc. Fifteenth Annual ACM-SIAM Symposium on Discrete Algorithms, SODA, New Orleans, Louisiana, USA, pp. 541–549 (January 2004)
26. Wang, M., Ailamaki, A., Faloutsos, C.: Capturing the spatio-temporal behavior of real traffic data. In: IFIP Int. Symp. on Computer Performance Modeling, Measurement and Evaluation, Rome, Italy (September 2002)
27. Wang, M., Papadimitriou, S., Madhyastha, T., Faloutsos, C., Change, N.H.: Data mining meets performance evaluation: Fast algorithms for modeling bursty traffic. In: Proc. Int. Conf. on Data Engineering, pp. 507–516 (February 2002)
28. Gedik, B., Wu, K.-L., Yu, P.S., Liu, L.: Adaptive load shedding for windowed stream joins. In: Proc. Int. Conf. on Information and Knowledge Management, CIKM, Bremen, Germany, pp. 171–178 (November 2005)

Reducing the Semantic Heterogeneity of Unstructured P2P Systems: A Contribution Based on a Dissemination Protocol

Thomas Cerqueus[1], Sylvie Cazalens[1,2,*], and Philippe Lamarre[3]

[1] LINA - UMR 6241 – University of Nantes
2, rue de la Houssinière
44322 Nantes Cedex 3 - France
`thomas.cerqueus@univ-nantes.fr`
[2] INRIA Sophia Antipolis - Méditerranée, Montpellier Center
161, rue Ada
34095 Montpellier Cedex 5 - France
`sylvie.cazalens@univ-nantes.fr`
[3] LIRIS - CNRS - UMR 5205, F69621 – University of Lyon
Campus de la Doua.
69622 Villeurbanne - France
`philippe.lamarre@liris.cnrs.fr`

Abstract. In resource sharing P2P systems with autonomous participants, each peer is free to use the ontology with which it annotates its resources. Semantic heterogeneity occurs when the peers do not use the same ontology. For example, a contributing peer A (e.g. a doctor) may annotate its photos, diagrams, data sets with some ontology of its own, while peer B (e.g. a genetician) uses another one. In order to answer a query issued in the system, peers need to know alignments that state correspondences between entities of two ontologies. Assuming that each peer has some partial initial knowledge of some alignments, we focus on correspondences sharing between the peers as a means to learn additional correspondences. We first provide several measures of semantic heterogeneity that enable to draw a semantic picture of the system and to evaluate the efficiency of protocols independently of query evaluation. We propose CorDis, a gossip-based protocol that disseminates the correspondences that the peers want to share in the system. To overcome the peers' storage limitations, we propose to consider a history of past queries and to favor the correspondences involving frequently used entities. We study several policies that a peer may adopt in case of inconsistency *i.e.* when shared correspondences conflict with its own knowledge. We conduct experiments with a set of 93 ontologies actively used in the biomedical domain. We evaluate the CorDis protocol with respect to the proposed measures of semantic heterogeneity and show its good behavior for decreasing them in several contexts.

* The second author thanks the LIRIS where she currently is an invited researcher, in the DRIM team.

A. Hameurlain et al. (Eds.): TLDKS VII, LNCS 7720, pp. 62–95, 2012.

1 Introduction

Peer-to-peer (P2P) systems have proved useful for sharing resources at large scale. In addition to their scalability and dynamicity properties, they enable the peers' autonomy and decentralized control. Each peer may be viewed as a resource provider which can decide both the resources it shares and the way it characterizes them. We focus on semantic unstructured P2P applications where each peer uses an ontology to represent its resources. An ontology is often presented as the explicit specification of a conceptualization of some domain [25]. It provides a controlled vocabulary to model a domain in terms of *entities*, namely concepts, concept properties and relations between concepts (less-general-than relation and possibly many others). The entities may be used to annotate the peer's resources, e.g. its photos, textual documents or data sets. Because the peers are free to choose the ontology that best fits their needs, it is unlikely that they all use the same one.

The use of different ontologies in the system results in the inability of some peers to precisely understand other peers' queries, thus providing poorly relevant resources in response. To overcome this problem, Peer Data Management Systems (PDMS) like [26,16,11] generally assume that neighbor peers know mappings (or alignments) between their schemas, *i.e.* a set of correspondences between the schema entities. The proposed systems consider relational or XML data, retrieved with a corresponding language like XQuery. Schema mappings expression and query reformulation are the main focus of [26,11] and nice results have been obtained. The authors of [1,16] in addition consider query routing in the peer network. Basically, a peer that receives a query treats it locally, then translates and forwards it to a neighbor peer whenever it knows some mapping between its schema and its neighbor's schema. This approach aims at exploiting the transitivity of schema mappings along paths in the peer network and its feasibility has been studied in [15]. It requires each peer a limited knowledge of schema mappings. We consider another approach where the peers learn more mappings during the system lifetime, by simple exchange of this knowledge between peers. Thus, a peer is more likely to answer an original query rather than a translated one.

In this work, our goal is to define an algorithm that enables the peers of a semantic P2P system to learn additional correspondences, with the following assumptions. First, the ontologies that are used to annotate the peers' resources are not populated, *i.e.*, we do not consider instances. To retrieve resources, a query is composed of, or annotated by, several entities that may be weighted or not. Second, in the query evaluation protocol, each peer translates the incoming query for its local needs but forwards the original query to its neighbors. The big advantage is that it lets the peer master its own translation choices with respect to the initial query. Each peer regularly uses some alignment process between its ontology and its neighbors' [22]. This results in an initial set of known correspondences. For the peer to learn additional correspondences, in particular with ontologies it does not know yet, we do not assume the existence of some peer that would be specialized in processing alignments and in maintaining a huge central repository. To our mind, this hypothesis would not fit the general P2P philosophy, primarily based on decentralization. Instead, we aim at defining a protocol that enables the peers to *share correspondences independently of query evaluation.* This raises several questions.

A first issue concerns characterizing what we call *semantic heterogeneity*, as a means to better describe the semantic picture of a system. Indeed, to our view, assuming that the correspondences sharing protocol runs in parallel but independently of the query evaluation protocol requires being able to precisely determine the performances of each one. Of course, it is possible to evaluate the global capacity of the system to ensure semantic interoperability between the peers. This can be done with application dependent measures like for example the precision and recall measures in Information Retrieval [8]. However, nothing enables to precisely determine the contribution of the correspondences sharing algorithm, nor to define the initial difficulty of the problem. Intuitively, if there are few ontologies or if the peers know a lot of correspondences, *i.e.* when semantic heterogeneity is low, ensuring interoperability seems easier to resolve. However, to the extent of our knowledge, no definition of semantic heterogeneity exists. Based on the observation that the concept of heterogeneity has several dimensions (or facets), we propose a typology of measures and several definitions to capture the different facets. We use some of these measures to evaluate the correspondences sharing protocol that we propose.

A second issue is the definition of a generic algorithm that enables the peers to share correspondences, with the aim of reducing the semantic heterogeneity of the P2P system along some dimensions. The idea is to make the peers share their knowledge by *disseminating* correspondences between entities of different ontologies. In order to implement dissemination of correspondences, we use a *gossiping* algorithm [31]: each peer regularly picks up some other peer for a two-way information exchange. In our case, each peer selects some correspondences to send to another peer. This latter also selects correspondences and sends them to the former. After several rounds, correspondences disseminate across the system. The CoRDiS protocol that we propose is based on this idea. In addition, because peers generally have limited local storage capacity, a scoring function is used to order the correspondences and store the most relevant ones. Relevance is computed considering a history of the incoming queries. We propose to favor the correspondences that involve entities that appeared in recent queries, and, to some extent chosen during the tuning of the application parameters, those involving entities belonging to ontologies referred to in recent queries. The scores of the correspondences are regularly updated, so that the CoRDiS protocol adapts the information exchange to the current queries.

A third issue concerns the effects of the presence of inconsistencies in a peer's knowledge base. Indeed, the first version of the CoRDiS protocol assumes that the peers always send correct correspondences and that they never conflict. Not considering this hypothesis leads to a twofold reflexion that encompasses both the way each peer deals with inconsistency and the way we take inconsistency into account within our heterogeneity measures. In the field of Logic in general, of Description Logics in particular, many works focus on inconsistency and solutions have been proposed to detect inconsistencies, and to some extent, repair, *i.e.* rule out knowledge that seems to cause inconsistency [39,7,40]. Based on existing results, our goal is to discuss several policies that a peer may apply in case of inconsistency, from a local point of view and a social one. Our second goal is to discuss the relevance of our heterogeneity measures in the presence of inconsistencies. We comment two possible options: either introduce measures that reflect how inconsistent the system is or adapt some heterogeneity measures.

Finally, a recurring issue is the way to conduct the experiments. We used the PeerSim simulator [35] to generate P2P systems as directed graphs and we implemented the CorDis protocol in Java. In such experiments, the real challenge is the choice of the ontologies. A common approach consists in choosing a given ontology and to randomly operate some changes on it to obtain other ontologies. Then, the ease with which the correspondences are obtained depends on the type of changes performed. Despite its systematic aspect, one may argue this does not correspond to a real-world ontologies set. This is why we consider 93 ontologies that are actively used in biomedical communities and accessible through the BioPortal Web services [23]. We evaluate the CorDis protocol with respect to the proposed measures of semantic heterogeneity and show its good behavior, as after a number of cycles that depends on the complexity of the initial situation, it lowers some facets of heterogeneity.

1.1   Illustrative Example

We provide a simple example that illustrates our hypotheses and the benefits of sharing knowledge of correspondences. We consider the semantic P2P system of Fig. 1. We assume that peers p_1, p_4 and p_8 use the same ontology o_1, but they do not know each others because they are very far (but not disconnected). There is no central repository of all correspondences between all the ontologies used in the system. Peers p_1, p_4 and p_8 have processed a matching algorithm with their respective neighbors. As a consequence, p_1 and p_8 know some correspondences between o_1 and o_2. On the contrary, peer p_4 knows no correspondences between o_1 and o_2.

For the ease of presentation, assume the query is flooded, although we do not use such algorithm in our research. When a peer receives a query, it translates it locally using the correspondences it knows and forwards the original query to its neighbors. This ensures the initial query is proposed to the peers which consequently can master their own translation choices. Let us assume that peer p_2 issues a query, characterized by the set of concepts c_2, c_2' and c_2'' of its ontology (o_2). When p_1 receives the query, based on the correspondences it knows, it performs some translation. Later, when p_8 receives the query, it may also be able to translate p_2's query. However, when p_2's query reaches it, p_4 has no knowledge of any correspondences between o_1 and o_2. Nor do its neighbors. Thus it cannot answer the query. If knowledge of correspondences between between o_1 and o_2 had been shared by both p_1 and p_8, either totally or partially depending on their wish, then p_4 would have been able to translate p_2's query. This paper

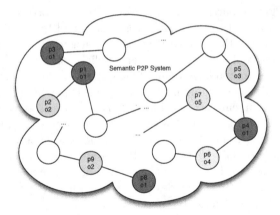

Fig. 1. Example of semantic P2P system

considers (i) sharing the correspondences by disseminating them across the system with a gossiping algorithm and (ii) studying the impact of the algorithm on the decrease of heterogeneity.

Peers p_1 and p_8 may have computed the alignments differently. So they may not have the same correspondences. For example, p_1 may know correspondences for all the three concepts c_2, c_2' and c_2'', while p_8 only knows a correspondence for c_2'. If p_1 shares its knowledge, then in the course of time p_8 learns it and is able to better translate the query. Notice that it may also happen that p_1 and p_8 do not agree on the correspondences concerning c_2', *i.e.* their respective knowledge may conflict. This paper focuses on several policies that peers may use in such a case.

Finally, notice that the problem would be different if p_6 and p_7 used ontology o_2 and p_4 knew many correspondences between o_1 and o_2. In other words, to better evaluate the efficiency of protocols, it is necessary to be able to draw the semantic picture of the initial system and its evolution. We focus on proposing measures to reflect different facets of the semantic heterogeneity of a system.

1.2   Contributions

This article is a revised and extended version of [12]. It brings several contributions. We introduce background knowledge used in the following (section 2) and the model in a more detailed way. We first consider several definitions of semantic heterogeneity measures of [12]. As they correspond to different facets of this notion (section 3), we in addition propose a typology of these measures. Second, we detail the algorithms of the CORDIS gossip-based protocol to disseminate correspondences across the system (section 4) introduced in [12]. The protocol considers a history of queries to score the correspondences. Thus it ensures some flexibility with respect to current queries. The work in [12] assumes the gossiped correspondences never result in inconsistence of peers' knowledge bases. Thus, in a third step, we define several policies that can be used by peers in case their knowledge base becomes inconsistent. Moreover, we provide ways to simultaneously take into account (*i.e.* measure) inconsistency and semantic heterogeneity (section 5). Fourth, we report on several experiments conducted with the PeerSim simulator. We switched from the small OntoFarm data set used in [12] to a bigger data set consisting of 93 ontologies from the NCBO Bioportal which are currently used by the biomedical community (section 6). The CORDIS protocol is evaluated with respect to the proposed measures of semantic heterogeneity. The results show that CORDIS significantly reduces several facets of heterogeneity while the network traffic and the storage space are bounded. This work builds on previous results concerning ontology mapping, ontology distances, gossiping algorithms and inconsistency management to define the CORDIS protocol that lowers some facets of semantic heterogeneity. However, it does not have any equivalent among the previously proposed solutions to improve semantic interoperability (section 7).

2   Background

2.1   Ontologies and Alignments

An ontology is often presented as the explicit specification of a conceptualization of some domain [25]. There are different types of ontologies with different expressiveness levels but they all provide a controlled vocabulary to model a domain using concepts and their properties and relations between these concepts. In this work, we consider that an ontology is composed of a set of concepts, a set of relations (among which a *less-general-than* relation) and a set of properties (assigned to concepts). Unlike some other works, we do not consider the instances of the concepts, *i.e.* we do not consider what is called a populated ontology.

For example, the left hand side of Fig. 2 shows a part of an ontology about plants where the concept *Flower*, less general than the concept *Thing*, is composed of concepts *Sepal*, *Petal*, *Corolla* and *Calyx*. It has three properties *color*, *size* and *odor*. Finally, concepts *Orchid* and *Edelweiss* are less general than the concept *Flower*.

The standard language OWL defined by the World Wide Web Consortium (W3C) [6] allows to represent ontologies by defining *classes*, *datatype properties* and *object properties*. Version 2 of OWL defines several profiles that correspond to different expressiveness levels and typical use cases. They rely on different description logics [7]. Except the fact that we do not consider instances, most of the work presented in this paper is independent of the type of profile used in the application. In section 5, we define the notion of inconsistency of a set of axioms in a general way, as the absence of model for this set.

We assume that each ontology is uniquely identified by a URI. Thus two ontologies are equal if and only if their URIs are the same. When there are different ontologies about the same domain, or close domains, it is useful to know whether two entities of two different ontologies refer to the same entity. An alignment process aims at identifying a set of correspondences between the entities of two ontologies [22,10].

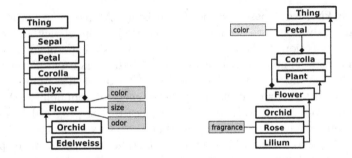

Fig. 2. Two ontologies o_1 and o_2 composed of concepts, properties and relations

Definition 1 (Correspondence). *A correspondence between two ontologies o and o' is a 4-tuple* $\langle e, e', r, n \rangle$ *such that e is an entity from o, e' is an entity from o',* $r \in \{\sqsubseteq, \equiv, \sqsupseteq, \perp\}$ *is a relation between e and e' and* $n \in [0, 1]$ *is a confidence value representing how much the correspondence is trustworthy.*

In this definition, the symbols $\sqsubseteq$, $\equiv$, $\sqsupseteq$ and $\perp$ respectively mean *less general than, equivalent to, more general than,* and *disjoint from.* Notice that an alignment is not necessarily perfect in the sense that some correct correspondences may be missing and others may be incorrect. An alignment between ontologies of Fig. 2 could contain correspondences presented on Table 1.

Table 1. A set of correspondences

Entity 1	Entity 2	Relation	Confidence
o_1:*Thing*	o_2:*Thing*	$\equiv$	1
o_1:*Flower*	o_2:*Flower*	$\equiv$	1
o_1:*odour*	o_2:*fragrance*	$\equiv$	0.8
o_1:*Edelweiss*	o_2:*Flower*	$\sqsubseteq$	0.9

2.2 Peer-to-Peer Systems

Peer-to-peer (P2P) systems have been proposed as an alternative to the client-server model, and became very popular for data sharing in the early 2000s [42]. In such systems, each peer acts as a client and a server at the same time. None of them is privileged. We focus on unstructured P2P systems [21] because they have good properties. First, they are fault-tolerant: the failure of a peer does not endanger the system. Second, they are scalable: the number of participants in the system does not damage performances, contrary to the bottleneck problem encountered with the client-server model. Finally they support peers' dynamicity, *i.e.* the fact peers may leave/join the system at any time. Popular implementations of unstructured P2P systems are Gnutella 0.4[1] and KaZaA[2].

---

[1] http://www.gnutellaforums.com

[2] http://www.kazaa.com

In unstructured P2P systems, peers maintain connections with other peers by storing their addresses in routing tables: the routing table of a peer p is noted $table(p)$. We assume that each peer p has a unique identifier, denoted by $id(p)$, sufficient to contact it.

Definition 2 (Unstructured P2P system). *An unstructured P2P system is defined by a graph $\mathcal{S} = \langle \mathcal{P}, \mathcal{N} \rangle$, where $\mathcal{P}$ is a set of peers and $\mathcal{N}$ represents a neighborhood relation defined by: $\mathcal{N} = \{(p_i, p_j) \in \mathcal{P}^2 : p_j \in table(p_i)\}$.*

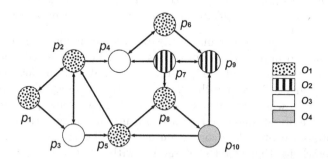

Fig. 3. Example of an unstructured P2P system where peers use different ontologies

Basically unstructured systems have no centralized information. It ensures the properties stated above but has the disadvantage of making data retrieval tough, because no centralized index is maintained, contrary to a structured system with a Distributed Hash Table (DHT) for example. Ideally queries should reach all the peers of the system to obtain relevant data, but as the number of peers may be important, most routing methods limit the flooding to a given number of hops: this number is called the TTL (Time To Live) of the query [3,4]. In this context, the neighborhood of a peer is of a first importance. The neighborhood of p within a radius r, denoted by $\mathcal{N}_p^r$, is defined as the set of peers accessible from p with l hops, where $1 \leqslant l \leqslant r$. We consider that p does not belong to its own neighborhood. In the system presented in Fig. 3, $\mathcal{N}_{p_1}^2$ is composed of p_2, p_3, p_4 and p_5.

2.3 Peers' Knowledge

We assume that each peer uses its own conceptualization of an application domain. The used ontology might be some reference ontology, a personalized extract of some reference ontology [24] or a specifically designed ontology. Of course, some peers may use the same ontology. We model the use of an ontology by a peer as a mapping from the set of peers to a set of ontologies.

Definition 3 (Peer-to-ontology mapping). *Given a P2P system $\mathcal{S} = \langle \mathcal{P}, \mathcal{N} \rangle$ and a set of ontologies $\mathcal{O}$, a peer-to-ontology mapping is a function $\mu : \mathcal{P} \rightarrow \mathcal{O}$, mapping each peer to one ontology.*

Each peer knows a set of correspondences used to bridge the gap between different ontologies, at least to a certain extent. The correspondences may be useful
to translate incoming queries received by a peer p with ontology o. Indeed, if p
receives a query q expressed with concepts of another ontology o', it has to know
some correspondences between o and o' to be able to (partially) translate and
treat it. We model a peer's knowledge of correspondences by a mapping from
the set of peers to a set a correspondences.

Definition 4 (Peer-to-correspondences mapping). *Given a P2P system*
$\mathcal{S} = \langle \mathcal{P}, \mathcal{N} \rangle$ *and a set of correspondences* $\mathcal{C}$, *a peer-to-correspondences mapping*
is a function $\kappa : \mathcal{P} \to 2^{\mathcal{C}}$, *mapping each peer to a set of correspondences.*

The way each peer internally manages the evolution of its ontology is a problem
by itself and lies far beyond the scope of this paper. The interested reader should
consult an overview paper like [27]. We assume that the ontology choice and
evolution is each peer's responsibility: It uses the ontology that best fits its
needs and decides when it is necessary to add/remove some parts to/from it.
In particular, we do not describe how it might use incoming correspondences to
make the ontology evolve. As a consequence, its set of correspondences is mainly
viewed as knowledge that enables interacting with other peers.

  We do not make any assumption about the way queries are transmitted in
the system, but we consider that they are unchanged during the propagation:
each peer receives and forwards the same query. It is responsible to internally
translate it if necessary.

2.4   Disparity between Peers

There are several reasons that make peers differ. Here, we only consider those
linked to semantics. Beyond the ontology, it is probably the whole way the peer
confronts the application domain that should be modeled. The way it evaluates
the similarity of two concepts in its ontology could be taken into account [13].
Indeed, peers may use the same ontology but differ by the similarity value they
give to a couple of concepts. For example, for a florist, *rose, orchid, lilium* may
be the concepts which are the most similar to *flower*, while for a biologist, may
be *plant* is the most similar. This is why we distinguish the disparity between
peers and the distance between ontologies, although obviously, the latter helps
characterizing the former.

Definition 5 (Disparity function). *A disparity function* $d : \mathcal{P} \times \mathcal{P} \to [0,1]$ *is*
a function that assigns a real value in $[0,1]$ *to a couple* $\langle p, p' \rangle$ *representing how*
much p' *differs from* p. *It satisfies the minimality property:* $\forall p \in \mathcal{P}, d(p,p) = 0$,
but we do not assume it is a mathematical distance.

Among the metrics used to estimate a distance between two ontologies, we point
out:

  – Average linkage dissimilarity [18]: given a dissimilarity measure between con-
    cepts of two ontologies, it computes the average dissimilarity between each

pair of concepts. The dissimilarity measure between concepts can be a label-based distance, a structural distance, etc.

- Taxonomy Overlap Average similarity [34]: it compares two ontologies' hierarchies to estimate their similarity.
- Largest covering preservation pre-similarity [19]: it considers the number of entities that can be translated through a path of alignments. In this case, the similarity (which is the inverse of the disparity) between two peers' ontologies depends on the successive translations. In this work, authors propose several measures of similarity in the alignment space [19]. It is close to our context if we consider that alignments are known by all peers.

In this work we consider that the disparity of a peer p' w.r.t. to another peer p depends on the capacity of p' to understand queries issued by p. This way, it is clear that the more disparate are peers, the harder it is to communicate.

The following equation defines a simple measure of disparity between two peers:

$$d_1(p, p') = \begin{cases} 0 \text{ if } \mu(p) = \mu(p') \\ 1 \text{ otherwise} \end{cases} \tag{1}$$

The measure may equal 0 (when both peers use the same ontology), or 1 (when peers do not use the same ontology). The advantage of this measure is that peers are easily able to compute it by themselves (they only have to exchange the URI of their ontologies). Nevertheless it does not capture the fact that two different ontologies may be very similar or very dissimilar. To overcome this problem, we propose another measure that also considers the number of correspondences between peers' ontologies:

$$d_2(p, p') = \begin{cases} 0 & \text{if } \mu(p) = \mu(p') \\ \frac{|\{e \in \mu(p) : \nexists \langle e, e', r, n \rangle \in \kappa(p') \text{ s.t. } e' \in \mu(p')\}|}{|\{e \in \mu(p)\}|} & \text{otherwise} \end{cases} \tag{2}$$

where $r \in \{\sqsubseteq, \equiv, \sqsupseteq, \perp\}$ and n is greater than a given threshold. In this equation, $\kappa(p')$ represents the set of correspondences known by p' between its ontology and p's ontology. This measure is more refined than d_1 because when ontologies of p and p' are different, it considers the set of correspondences existing between them.

3   Definition of Semantic Heterogeneity Measures

Through the analysis of different scenarios [13], we have shown that quantifying heterogeneity, although a bit reductive, is still very informative and that several measures should be defined to reflect the different facets of the semantic heterogeneity of a system. It is indeed very difficult to capture as diverse characteristics as the number of ontologies on use, the disparity between peers or the organization of the system into a single measure. We mathematically define the notion of heterogeneity measure, before describing our typology and proposing several measures. Notice that our main concern is to be able to characterize the global semantic picture of a system for experimental studies. Hence, it is coherent to assume entire knowledge of the system. Of course, in a real system a peer only has a limited knowledge and some of the proposed measures do not apply.

Definition 6 (Semantic heterogeneity measure). *Let $\mathcal{M}$ be a set of models $m = \langle \mathcal{S}, \mathcal{O}, \mu, d \rangle$ where $\mathcal{S}$ is a P2P system, $\mathcal{O}$ is a set of ontologies, μ is a peer-to-ontology mapping, and d is a disparity function between peers.*
A semantic heterogeneity measure is a function $\mathcal{H} : \mathcal{M} \to [0, 1]$ such that:

- *$\mathcal{H}(m) = 0$ if $|\{o \in \mathcal{O} : \exists p \in \mathcal{P} \text{ s.t. } \mu(p) = o\}| = 1$ (minimality);*
- *$\mathcal{H}(m) = 1$ if $\forall p \neq p' \in \mathcal{P}, d(p, p') = 1$ (maximality).*

The conditions express that (i) homogeneity occurs when the same ontology is used by all the peers and that (ii) maximal heterogeneity occurs when all the disparity values between peers are maximal. Notice that correspondences known by peers are not explicitly mentioned in the model. Indeed they are implicitly considered in the disparity measure d.

Based on this model, we propose a typology of heterogeneity measures. The idea is to distinguish them according to the elements of the model that are considered in their definition. In our view, every measure should at least consider $\mathcal{P}$, $\mathcal{O}$ and μ which are the basic components of the model. Then, we differentiate the heterogeneity measures which are:

- *Structure aware/unaware*: An heterogeneity measure is structure aware if its definition considers the neighborhood relation $\mathcal{N}$ of the P2P system. Otherwise it is structure unaware.
- *Disparity aware/unaware*: An heterogeneity measure is disparity aware if its definition considers the disparity function d between peers. Otherwise, it is disparity unaware.

These two criteria can be combined, leading to four classes of heterogeneity measures. For each class, Table 2 enumerates the elements of the model that should be considered in the definition of a semantic heterogeneity measure of this class.

Table 2. Four classes of heterogeneity measures.

	Structure unaware	Structure aware
Disparity unaware	$\mathcal{P}, \mathcal{O}, \mu$	$\mathcal{P}, \mathcal{O}, \mu, \mathcal{N}$
Disparity aware	$\mathcal{P}, \mathcal{O}, \mu, d$	$\mathcal{P}, \mathcal{O}, \mu, \mathcal{N}, d$

Depending on the system to evaluate, several measures of each class may be useful to capture all the aspects of heterogeneity. In the remaining of this section, we propose measures which are general enough to be used in many application domains while still being meaningful.

3.1 Structure Unaware Measures

Disparity Unaware Measure. Notion of diversity is commonly used to measure the heterogeneity of a population (e.g. in biology). Richness partly characterizes the diversity of a population. In our context it depends on the number of different ontologies used in the system. If all the peers use the same ontology, then the system is completely homogeneous. By cons, the more ontologies there are, the more heterogeneous the system is. This idea can be expressed by the following measure:

$$\mathcal{H}_{Rich}(m) = \frac{|o_\mathcal{S}| - 1}{|\mathcal{P}| - 1} \tag{3}$$

where $|o_\mathcal{S}|$ is the number of different ontologies used in the system $\mathcal{S}$, and $|\mathcal{P}|$ the number of peers. In the system presented on Fig. 3, four different ontologies are used by the ten participants: $\mathcal{H}_{Rich}(m) = \frac{4-1}{10-1} = 0.33$. A richness value equal to 0 means that heterogeneity is null. A value equal to 1 means that heterogeneity is total: alignments are needed between each pair of participants to communicate.

Disparity Aware Measure. On top of determining diversity, it is interesting to take into account the disparity between peers of the system. Indeed diversity measures do not make any difference between a system $\mathcal{S}_1$ with n peers between which similarities are important, and a system $\mathcal{S}_2$ with n peers between which similarities are weak.

We propose to consider disparity between peers rather than only consider the ontologies they use. If the disparity between peers is globally important, it means that peers have important knowledge differences. The more different their knowledge, the harder to communicate (*i.e.* answering queries). Indeed an important loss of information will occur during query translation. As we do not take into account the system topology, we consider the disparity between each pair of peers:

$$\mathcal{H}_{Disp}(m) = \frac{1}{|\mathcal{P}|^2 - |\mathcal{P}|} \sum_{p_i \neq p_j \in \mathcal{P}} d(p_i, p_j) \tag{4}$$

The $\mathcal{H}_{Disp}$ measure determines if peers are globally disparate from each other. The factor $\frac{1}{|\mathcal{P}|^2 - |\mathcal{P}|}$ is used to normalize to measure in $[0, 1]$.

3.2 Structure Aware Measures

Disparity Unaware Measures. The neighborhood of a peer in an unstructured P2P system is of prime importance because it represents the set of its privileged interlocutors. Thus we think it is crucial to focus on peers neighborhoods, considering the topology of the system. First of all we can study a peer's neighborhood numbering the peers that cannot understand it directly (*i.e.* without using any alignment):

$$\mathcal{H}^r_{Rap}(m,p) = \frac{|\{p_i \in \mathcal{N}^r_p : \mu(p_i) \neq \mu(p)\}|}{|\mathcal{N}^r_p|} \tag{5}$$

This measure gives basic information about a peer's neighborhood and could be computed by the peer itself. Indeed, it just requires to be able to determine if another peer uses the same ontology.

Example 1. On the example of Fig. 3, the neighborhood of p_1 (with $r = 2$) is composed of p_2, p_3, p_4 and p_5. As p_3 and p_4 do not use the same ontology as p_1, we find that $\mathcal{H}^2_{Rap}(m, p_1) = \frac{2}{4} = 0.5$.

In order to get a heterogeneity measure (as it is defined in definition 6), we apply $\mathcal{H}_{Rap}$ on each peer of a system:

$$\mathcal{H}^r_{RapAvg}(m) = \frac{1}{|\mathcal{P}|} \sum_{p \in \mathcal{P}} \mathcal{H}^r_{Rap}(m,p) \tag{6}$$

Disparity Aware Measures. The fact that two peers do use the same ontology does not induce that they can not communicate together. So we can refine the previous measure by considering a disparity measure:

$$\mathcal{H}^r_{Dap}(m,p) = \frac{1}{|\mathcal{N}^r_p|} \sum_{p_i \in \mathcal{N}^r_p} d(p,p_i) \tag{7}$$

Example 2. *As p_1, p_2 and p_5 use the same ontology (o_1), we have $d(p_1, p_2) = d(p_1, p_5) = 0$. As p_3 and p_4 use the same ontology o_3, we find that:*

$$\mathcal{H}^2_{Dap}(m, p_1) = \frac{1}{4}[d(p_1, p_3) + d(p_1, p_4)]$$

Again a global measure can be obtained:

$$\mathcal{H}^r_{DapAvg}(m) = \frac{1}{|\mathcal{P}|} \sum_{p \in \mathcal{P}} \mathcal{H}^r_{Dap}(m,p) \tag{8}$$

If $\mathcal{H}_{DapAvg}$'s value is low, peers are globally close to their neighbors: each peer is surrounded by peers able to "understand" it.

Proposition 1. *All the measures introduced in this section satisfy both properties of minimality and maximality (proof is trivial).*

4   The CorDis Protocol

In this section we present CORDIS, a gossip-based protocol. We first explain what we mean by *gossip-based protocol*. Then we describe the specificities of CORDIS.

4.1   Gossiping Protocols

In gossip-based protocols [31], each peer consists of two threads: an active and a passive one. The active thread is used to initiate communications with another peer. Thus, each peer regularly contacts another peer to exchange information. When a peer is contacted by another one (through the passive thread), the former has to answer by sending some information. Thus, both peers treat the received information. In these algorithms, peers have to process three crucial tasks: peer selection, data selection and data processing. In P2P systems, the peer selection can be ensured by a peer sampling service. This service allows peers to uniformaly and randomly select another peer [30]. The other two tasks depend on the target. If the protocol is used to disseminate information, the data selection consists in selecting some information, and the data processing consists in storing the received information.

4.2   Main Principles of CorDis

The purpose of CorDis is to disseminate correspondences over the network to share those known by some peers but ignored by others. The underlying goal is to reduce the disparity between peers, and thus to reduce the semantic heterogeneity related to these disparities (see equations (4) and (8)).

Algorithm 1: Instructions performed in the active thread.

```
// Executed by a peer p
1 while true do
2 │   p' ← selectPeer() ;                    // ensured by a peer sampling service.
3 │   msg ← selectCorrespondences(m_max) ;
4 │   sendTo(p', msg) ;
5 │   receiveFrom(p', msg') ;
6 │   storeCorrespondences(msg') ;
7 └   wait(θ) ;                              // θ is the period.
```

Algorithm 2: Instructions performed in the passive thread.

```
// Executed by a peer p'
1 receiveFromAny(p, msg) ;
2 msg' ← selectCorrespondences(m_max) ;
3 sendTo(p, msg') ;
4 storeCorrespondences(msg) ;
```

When the process starts, each peer p knows some correspondences. This set, noted $\kappa_0(p)$, should always be recorded by the peer because they is considered as reliable. Basically these correspondences are computed between p and its neighbors. They are of a prime importance for p because they involve its own ontology:

they are called o_p-correspondences. The purpose of dissemination is that each peer learns additional correspondences that might be useful to it to translate the queries it receives into its own ontology. We disseminate the correspondences by gossiping: Each peer p regurlarly initiates an exchange of correspondences with another peer p'. It selects some correspondences it knows and sends them to p'. In turn, p' chooses among the correspondences it stores and send them to p. Algorithms 1 and 2 summarize this protocol.

4.3   Storage of Correspondences

Each peer must store the correspondences it has been informed of in some cache, of limited size, thus preventing the peer from storing all the correspondences. Choice of the correspondences to keep is obtained by a scoring function which enables to order the correspondences: only the best ones are kept. In theory, the scoring function could be specific to each peer. Here we propose that each of them consider a history of the received queries.

A **history** of received queries is made of two lists $\mathcal{L}_1$ and $\mathcal{L}_2$. List $\mathcal{L}_1$ contains the entities used in the last received queries, while $\mathcal{L}_2$ contains the ontologies used to express the last received queries. Notice that an item can appear several time in a list if it has been involved in several queries. In order to limit the volume of stored data, we limit the size of both lists: $\mathcal{L}_1$ contains at most k items, and $\mathcal{L}_2$ contains at most k'. They are managed in a FIFO fashion.

Example 3. *After a peer p receives queries q_1, q_2, q_3 and q_4 respectively expressed with entities of ontology o_1, o_3, o_1 and o_2 such that:*

- *q_1 is composed of e_1,*
- *q_2 is composed of e_3,*
- *q_3 is composed of e'_1 and e''_1,*
- *q_4 is composed of e_2,*

its history is made of: $\mathcal{L}_1 = [e_1, e_3, e'_1, e''_1, e_2]$, $\mathcal{L}_2 = [o_1, o_3, o_1, o_1, o_2]$.

The intuition of the scoring function is that peers favor the correspondences that might be useful for translating queries (it can be useful locally, or for others).

Definition 7 (Scoring function). *Given a set of correspondences $\mathcal{C}$, we define the scoring function of a peer $sc : \mathcal{C} \to [0, 1]$ as:*

$$sc(\langle e, e', r, n \rangle) = \omega \cdot [f_1(e) + f_1(e')] + (1 - \omega) \cdot [f_2(o) + f_2(o')] \qquad (9)$$

where $e \in o$, $e' \in o'$, and f_1 (resp. f_2) measures the frequency of occurence of an element in $\mathcal{L}_1$ (resp. $\mathcal{L}_2$).

The coefficient $\omega \in [0, 1]$ is used to give more or less importance to a correspondence involving entities that do not appear in $\mathcal{L}_1$, but that belong to ontologies stored in $\mathcal{L}_2$.

Example 4. *Considering the history presented in example 3 and $\omega = 0.5$, the score of the correspondence $c = \langle e_1, e_2''', r, n \rangle$ is:*

$sc(c) = 0.5 \cdot [f_1(e_1) + f_1(e_2'') + f_2(o_1) + f_2(o_2)] = 0.5 \cdot [0.2 + 0 + 0.6 + 0.2] = 0.5,$
where e_1 and e_7' respectively belong to o and o'.

If the focus of interest of the queries changes, the scoring values of the correspondences will change, giving more importance to relevant correspondences. Scores are regularly calculated to take dynamicity into account.

Because the correspondences involving its own ontology (*i.e.* the o_p-correspondences) are of prime importance for the peer, we propose that each peer tries to store as much possible of them (or all of them if possible) in a specific **repository**, including $\kappa_0(p)$, distinct from the cache which is then devoted to the other correspondences. If the repository is too small for storing all the o_p-correspondences, the peer can use the scoring function to eliminate some of them. We denote by *repository(p)* the repository of a peer p, and by *cache(p)* its cache (respectively limited to r_{max} and c_{max} entries).

Algorithm 3: Instructions executed when the function storeCorrespondences is called (see algorithm 1, line 6 and algorithm 2, line 4).

```
// Executed by a peer p
input : A set set of correspondences.

// Correspondences are added in the repository or in the cache.
```
1 **for** $c = \langle e, e', r, n \rangle \in set$ **do**
2   **if** $e \in \mu(p) \vee e' \in \mu(p)$ **then**
       `// If c is an` o_p`-correspondence`
3        $repository(p) \leftarrow repository(p) \cup \{c\}$;
4   **else**
5        $cache(p) \leftarrow cache(p) \cup \{c\}$;

```
// Limitation of the size of the repository and the cache.
// r_max and c_max are application parameters.
```
6 **if** $|repository(p)| > r_{max}$ **then**
7     $sort(repository(p), sc)$;   `// sc is the scoring function`
8     $repository(p) \leftarrow \text{keepBest}(repository(p), r_{max})$;

9 **if** $|cache(p)| > c_{max}$ **then**
10     $sort(cache(p), sc)$;
11     $cache(p) \leftarrow \text{keepBest}(cache(p), c_{max})$;

When a peer p receives a message, it executes two main tasks. First it computes the score of the correspondences in *msg* and then merges them with its local data. It only consists in adding o_p-correspondences in *repository(p)* and the others in *cache(p)* and re-order the correspondences. If a correspondence is already stored, the newest score is used. Then, the best r_{max} (resp. c_{max}) correspondences are kept in the repository (resp. in the cache). Algorithm 3 explains this process. Best correspondences (according to the scoring function) are kept because they are potentially useful to translate queries.

4.4   Selection of Correspondences

When a peer has to send correspondences, it selects them from both the cache and the repository. We introduce the number $\pi \in [0, 1]$ to represent the ratio of correspondences to select in both sets. Thus, a peer randomly selects $[\pi \cdot m_{max}]$ correspondences in its repository, and $[(1 - \pi) \cdot m_{max}]$ in its cache. Algorithm 4 summarizes this process. Random selection is used to ensure that two correspondences of the repository (resp. the cache) have the same probability to be sent. Thus we aim to keep good properties of gossiping protocols.

Algorithm 4: Instructions executed when the function selectCorrespondences is called (see algorithm 1, line 3 and algorithm 2, line 2).

---

input : A maximal number max of correspondences to return.
output: A set set of correspondences.

1 $set \leftarrow$ selectFromRepository($\pi \cdot max$) ;   `// π is an application parameter.`
2 $set \leftarrow set \cup$ selectFromCache($(1 - \pi) \cdot max$) ;
3 **return** set ;

---

5   Dealing with Inconsistencies

So far, we assumed that the correspondences were computed by trustable peers and that they never introduced any inconsistency. Leaving this assumption, this section aims at (i) defining inconsistency in our context, (ii) defining possible peer's policies to deal with inconsistencies and (iii) discussing the ways to take into account both inconsistency and heterogeneity.

5.1   Context

Let us first recall that each peer's ontology plays a very specific role as it is used to annotate its resources, let it be manually or using some automatic indexing. The way the indexer proceeds is out of the scope of this paper, but it surely takes advantage of the rich meta-information provided by the ontology. Hence, we assume that the peer does not change its ontology automatically upon simple arrival of correspondences issued by other peers. In other words, each peer makes its ontology evolve in the course of a distinct process, which may require some human intervention. As a consequence, we assume that the peer ontology is consistent and that the gossiped correspondences are only used to translate queries.

However, intuitively, it may happen that a given set of correspondences together with the ontology leads to an inconsistency. For example, the peer's ontology states that concepts c and c' are disjoint, while the set of correspondences states or entails that c is included in c'. In that case, should the peer go on gossiping its correspondences? Should it still use the correspondences to translate incoming queries? Should it try to "repair" (debug) the set of correspondences that has become inconsistent with respect to its ontology? The peer may consider the confidence value of correspondences (see definition 1), or even some peers' confidence value in order to rule out some correspondences. But it cannot avoid cases where correspondences have the same confidence degree. In that case, it faces the general problem of detecting inconsistencies and repairing. This tough problem has been extensively studied in the field of Logic in general and in Description Logics (DL) in particular. Our objective is to use recent results to discuss several policies a peer may adopt in case of inconsistency.

5.2   Inconsistency of a Peer's Knowledge Base

Description Logics are knowledge representation formalisms to reason about the entities of a given domain [7]. They are the formal bases of ontological languages like OWL. Each description logic has its syntax and specific set-based semantics, which results in some expressiveness degree. A Knowledge Base (KB) is often presented as the union of a TBox and an ABox. The former contains all the axioms describing the application domain, while the latter contains all the assertions about individuals. As mentioned in section 2.1, we do not consider populated ontologies to annotate the resources. Hence we focus on TBoxes only. In logic $\mathcal{ALC}$ and more expressive ones, a set of axioms is consistent if and only if it has a model.

Without any intention to change the definition of inconsistency, it is interesting to more closely analyze where inconsistency of a peer's knowledge base may come from. Let o be the peer's ontology. We denote by $Ax(o)$ the set of axioms describing ontology o, and by SC the set of axioms representing the stored correspondences, $i.e.$ belonging to either the repository or the cache (see section 4). Then, the peer's knowledge base is $Ax(o) \cup SC$.

Definition 8 (Inconsistency of peer's knowledge base). *A peer's knowledge base is inconsistent if $Ax(o) \cup SC$ has no model.*

Depending on the logic used, consistency checking is a reasoning task that may be complex. For several logics, including $\mathcal{ALC}$, tableaux methods have been defined and most reasoners can detect inconsistencies, like FaCT++ [43] or Pellet [41] which are actively used. Hence, we assume that each peer can use a reasoner to check the consistency of $Ax(o) \cup SC$ (upon every arrival of a new correspondence or at a lower frequency).

Notice that, by assumption, $Ax(o)$ is maintained consistent by the peer. The intrinsic consistency of SC depends on the types of correspondences that are gossiped. Indeed, if relations like disjointness or non-inclusion are allowed, then SC

may itself be inconsistent. However, we think that this type of correspondences does not directly help a query translation process and that, if systematically computed may overload the system network with not usable information. When only *positive* correspondences are gossiped (*i.e.* where no negation nor the bottom (impossible) concept are used to express them), the set SC is also consistent. Hence, inconsistency only happens when confronting the stored correspondences SC with the peer's ontology $Ax(o)$. This does not help to check for consistency, as the peer must still consider $Ax(o) \cup SC$. However a consequence is that, in case of consistency repair, the set of candidate axioms to be ruled out is included in SC.

Example 5. *Peer p makes a clear distinction between concepts Petal and Leaf i.e. Petal⊓Leaf ⊑ ⊥ belongs to $Ax(o)$. Peer p' has sent a correspondence stating that the concept Petal is a kind of Leaf, i.e. Petal ⊑ Leaf belongs to p's stored correspondences SC. In that case, there is no possible model for $Ax(o) \cup SC$. A repairing strategy would rule out Petal ⊑ Leaf.*

The fact that neither $Ax(o)$ nor SC can themselves be inconsistent restricts the possibilities. With no aim to be exhaustive, we list some typical cases involving two entities e and e' that raise inconsistency[3]:

- $Ax(o)$ states that e and e' are different and SC states they are equivalent.
- $Ax(o)$ states that e and e' are different with e' less general than e whereas SC states that e is less general than e'.
- $Ax(o)$ states that e and e' are disjoint and SC states they are equivalent.
- $Ax(o)$ states that e and e' are disjoint while SC states e is less general than e'.

5.3   Possible Policies of Peers in Case of Inconsistencies

We distinguish a peer's *local behavior* from its *social behavior*. The local behavior includes the way it checks for inconsistency, the way it modifies its stored correspondences (because of a repairing strategy) and which correspondences it uses to translate the incoming queries. Its social behavior defines the way it forwards the correspondences. It is independent of its local behavior. Indeed, a first approach is to assume that the peer's opinion has no reason to prevail over others' opinion. Hence, even if it does locally change its set of correspondences, it may go on forwarding all the correspondences it receives. Another approach consists in letting the peer favor its point of view by avoiding to forward some, or all the correspondences it considers doubtful. We illustrate these observations with three possible policies.

Reckless Policy. A peer may consider that the inconsistency of $Ax(o) \cup SC$ is not really a problem, because $Ax(o)$ is kept consistent (remember this is our assumption). A peer uses a reckless policy if:

---

[3] For the ease of presentation, we assume the axiom sets are closed by deduction.

– it does not even check the consistency of $Ax(o) \cup SC$,
– it does not withdraw any correspondence from SC and keeps translating the incoming queries using any correspondence in SC,
– it forwards any correspondence in SC.

Intuitively, for disseminating correspondences, peers being equal, none of them makes its view prevail. The advantage is that the peer saves computing resources because it does not have to check for consistency, every time a new correspondences comes (in the worst case). Of course, the drawback is that if a peer, intentionally or not, does send incorrect correspondences, they are disseminated in the whole system.

Cautious Policy. Another behavior consists in trying to only use the correspondences which do not seem to be concerned by any inconsistency. We classically introduce the notion of minimal inconsistent set of axioms to describe this policy [28].

Definition 9 (Minimal inconsistent set). *A minimal inconsistent set MIS of $Ax(o) \cup SC$ is any set of axioms $MIS \subseteq Ax(o) \cup SC$ such that MIS is inconsistent and for any $MIS' \subset MIS$, MIS' is consistent.*

Notice that an inconsistent set of axioms may have several minimal inconsistent subsets. Let us denote by S_{MIS} the set of minimal inconsistent subsets of $Ax(o) \cup SC$. Deriving from R. Reiter's work [39], it is possible to define algorithms to compute a *diagnosis*, *i.e.* a smallest set of axioms that needs to be removed or corrected to render a set of axioms consistent. In our case, the set of candidates for withdrawal is limited to a subset of SC.

Definition 10 (Doubtful correspondence). *A doubtful correspondence is any correspondence in SC that belongs to at least one set $MIS \in S_{MIS}$.*

Given these definitions, a peer uses a cautious policy if:

– it checks for consistency and computes S_{MIS} regularly,
– it uses no doubtful correspondence to translate the incoming queries and uses a repairing strategy that leads to the withdrawal of one or more correspondences from SC.
– it forwards no doubtful correspondence.

The advantage of this policy is that the peer only uses correspondences that are consistent with its own knowledge during the translation process. The social behavior is more debatable if we consider that the peer, having no special rights w.r.t. other peers, is not entitled to avoid the dissemination of some correspondences. However, if there are no malicious peers and if their views are compatible, it may be the best way to avoid propagating any "honest" mistake.

Policy Focused on Important Concepts. This proposal is based on two observations. First, some concepts seem more important than others in a peer's ontology. Second, it may be too costly to compute all the minimal inconsistent sets, even if the peer does not do it every time a new correspondence comes in. To define the notion of *important* concept, we propose to consider the following characteristics.

Key concepts. As ontologies may be very big, it can be useful to consider its *key concepts* as in [37]. The authors propose an algorithm to compute the key concepts based on several criteria linked to general cognitive principles, the structure of the ontology (coverage, density, ...) and the popularity with respect to some resource (e.g. Yahoo's search engine). Their purpose is to obtain a kind of ontology summary.

Representative concepts. As the ontology is used to annotate the peer's resources, some parts of the ontology may be more representative than others. For example, if the concept *phenotype* is used to annotate 50% of the documents with a weight over 0.7, this concept is more representative than *snow*, used only for 5% of the documents with weights under 0.3. Representativeness of concepts with respect to the documents can be obtained through an analysis of the index.

Popular concepts. One could say that a concept is popular if it is often queried, either directly or after translation. This requires the peer to maintain a richer version of its history of queries (see section 4), by counting, for each concept of its ontology, the number of times it appears in a translated query. The measure should be normalized by dividing this number by the total number of concept occurrences in the translated queries. Then a concept is popular if its counter is above some threshold.

Each peer can then define its important concepts, through a combination of the above criteria. Once they are identified, the idea is to avoid their problematic translation. Thus we propose to limit the diagnosis computation to the important concepts only and repair the set of correspondences consequently. We refer to the formal framework presented in [40], where the authors in addition propose a heuristically driven algorithm to approximate the set of diagnoses in case of "big" ontologies.

Finally, a peer uses a policy focused on important concepts if it maintains a list of the important concepts of its ontology and if:

- it computes a diagnosis for each important concept of its ontology and repairs the set SC,
- it uses the correspondences in SC to translate the incoming queries
- it forwards the correspondences in SC.

The advantage of this policy is that it ensures the peer to be coherent when translating queries involving its important concepts. The discussion about the social behavior (what to forward) is the same as before.

As a conclusion, there is no perfect policy and the most appropriate one should be chosen depending on the application. Indeed, if the risk of inconsistency is limited, it might not be worth requiring each peer to have a reasoner to check for consistency and/or to compute the minimal sets and adopt a repairing strategy.

5.4   Heterogeneity Measures and Inconsistencies

In this section, we briefly discuss the links between inconsistency and the different facets of heterogeneity and the way these links could be reflected by appropriate measures.

Disparity functions between peers might be impacted by the existence of doubtful correspondences, and so might be the overall heterogeneity values. There are basically two options. Either we keep the heterogeneity measures unchanged and we in addition introduce an inconsistency measure. This means that a low heterogeneity value (because of many correspondences between peers) should always be considered in the light of the inconsistency value of the system (there might be a lot of doubtful correspondences). Or we modify the definition of some heterogeneity measures to take inconsistency into account.

Introduction of an Inconsistency Measure. Several works have proposed measures of inconsistency for a knowledge base, either for classical logic or for Description Logics, for example based on para-consistent logic [33,32]. We adapt the definitions used in [28] to give very simple definitions, using minimal inconsistent sets $\mathcal{MIS}$. We refer to the cited papers for more sophisticated definitions.

The approach consists in evaluating the inconsistency introduced by a doubtful correspondence first, then in computing the overall inconsistency of the peer's knowledge base. Finally, the inconsistency of the whole system must be defined considering the inconsistency values of the peers.

Definition 11 (Inconsistency value of a doubtful correspondence). *Let corr be a doubtful correspondence. Then, the inconsistency value of corr, noted $IncV(corr)$, w.r.t. some knowledge base $\mathcal{A}x(o) \cup \mathcal{SC}$ is given by:*

$$IncV(corr) = \sum_{\mathcal{MIS} \,\in \mathcal{MIS}(corr)} \frac{1}{|\mathcal{MIS}|}$$

where $\mathcal{MIS}(corr) = \{\mathcal{MIS} \in S_{\mathcal{MIS}} : corr \in \mathcal{MIS}\}$

Instead of just counting the number of minimal inconsistent sets the doubtful formula belongs to, the above definition also considers the cardinality of the minimal inconsistent set. Intuitively, a conflicting correspondence has more impact if the number of correspondences in the minimal inconsistent set is low. For example, let us consider two correspondence $corr_1$ and $corr_2$. Correspondence $corr_1$ belongs to two minimal inconsistent sets which cardinality is respectively 10 and 5, while $corr_2$ belongs to two minimal inconsistent sets which cardinality is 2 and 4. Then, $IncV(corr_1) = 0.3$ and $IncV(corr_2) = 0.75$ which reflects the fact that $corr_2$, though belonging to the same number of inconsistent set, might be more conflicting because their cardinality is very low.

Given the inconsistency value of every correspondence, it is then possible to define an inconsistency value for a peer's knowledge base $\mathcal{A}x(o) \cup \mathcal{SC}$. Several choices are possible to aggregate the inconsistency values. In our view, it is important to reflect the worse case, *i.e.*, the maximal value.

Definition 12 (Inconsistency value of a peer). *Let $Ax(o) \cup SC$ be the knowledge base of some peer p. Let $\mathcal{D}$ be the set of doubtful correspondences, the inconsistency value of p, noted $IncV(p)$ is given by:*

$$IncV(p) = \max_{corr \in \mathcal{D}}(IncV(corr))$$

Then, for a given P2P system $\mathcal{S} = \langle \mathcal{P}, \mathcal{N} \rangle$ we consider the average of peers' inconsistency values, to reflect the extent to which the system is globally inconsistent. We do not consider the neighborhood relation $\mathcal{N}$ because, in our view, the fact that some peers are linked does not increase or decrease inconsistency.

In this approach, the system is still qualified by its heterogeneity measures (as defined in section 3) but given its inconsistency value.

Adapting the Heterogeneity Measures. Intuitively, we want to take into account the fact that some correspondences that a peer knows between its ontology and other peers' ontologies are doubtful. A simple idea is to modify the definition of disparity between two peers. For example, let us consider the equation (2) defined at the end of section 2. For some peer p, with respect to peer p', it considers the number of entities known by p for which p' knows a correspondence.

$$d_2(p, p') = \begin{cases} 0 & \text{if } \mu(p) = \mu(p') \\ \frac{|\{e \in \mu(p) : \nexists \langle e, e', r, n \rangle \in \kappa(p') \text{ s.t. } e' \in \mu(p')\}|}{|\{e \in \mu(p)\}|} & \text{otherwise} \end{cases} \quad (10)$$

In the previous equation, $\kappa(p')$ denotes the set of correspondences known by p'. We only have to modify this point, considering that $\kappa(p')$ only contains safe correspondences (*i.e.* correspondences that are not doubtful). We do not change the definition of the heterogeneity measures, but obviously, whenever the peers compute doubtful correspondences, the values of $\mathcal{H}_{Disp}$ and $\mathcal{H}_{DapAvg}$ get lower.

6   Experiments

Analytical and experimental evaluations of gossiping and dissemination algorithms have been conducted in several works, which results are summarized in [31]. The proposed experiments focus on the performances of the CORDIS protocol with respect to semantic heterogeneity. They both illustrate the intuitive meaningfulness of the proposed measures and the efficiency of the protocol. We run several simulations to study the evolution of some heterogeneity measures in function of (i) the semantic diversity (measured with $\mathcal{H}_{Rich}$), (ii) the number of peers in the system, (iii) the dynamicity of the system. In all the experiments we consider the reckless policy presented in section 5.3.

These experiments represent a first important step in the whole evaluation of the CORDIS protocol, which is difficult because of several factors. A first point is the data set. We preferred considering an actually used set of ontologies from the BioPortal [23], instead of using artificially created larger sets. This impacts the highest value of richness we can reach, and the number of peers in the system (if we want testing with the worse possible heterogeneity initial values).

Thus future experiments should check the intrinsic properties of the CORDIS protocol at an even larger scale. A second point is the difficult comparison with other works. Indeed, the specificity of our approach is to distinguish heterogeneity (that describes the semantic picture of the system) and interoperability (that describes the extent to which peers understand each others in the system). Hence our future software architecture would have two layers, one for heterogeneity decrease (including the CORDIS protocol) and another one for interoperability increase (including query evaluation protocols). To our knowledge, there is no work that makes such a difference and consequently that would provide detailed enough algorithms for heterogeneity decrease only. Thus, next evaluation steps will occur with experiments involving the whole software.

6.1   Material

Ontologies and Alignments. Using the BioPortal Web services [23] we obtained ontologies that are actively used in biomedical communities. BioPortal is an open repository of biomedical ontologies that provides access to ontologies developed in OWL, RDF, OBO. We translated OBO ontologies to OWL ontologies thanks to the ONTO-PERL API [5]. We obtained 149 ontologies workable with the OWL API [9]. We also retrieved 1417 alignments (sets of correspondences) between the ontologies (with a BioPortal Web service). We finally kept the 93 ontologies for which there exist one alignment with another ontology. The ontologies contain $1,100$ concepts in average representing an average volume of 1.7 MiB. The average number of correspondences is $28,027$. Given the number of concepts, and the number of correspondences, we can say that ontologies are very dissimilar: they hardly overlap.

Unstructured Semantic P2P System. We implemented the CORDIS protocol in Java and we used the PeerSim simulator [35] to generate P2P systems as directed graphs. We use the cycle-based model, so that we can make peers perform the Algorithm 1 regularly (at each cycle). The PeerSim simulator provides a random graph of peers: Each peer is linked to some others peers. These latters are used to initialized its view.

Besides, ontologies are randomly assigned to peers according to a Poisson distribution (with $\lambda = 5$). Some ontologies are used by more peers than others. We think that it is a realistic situation. Finally each peer initially knows some correspondences involving its own ontology. In average each peer knows only 20% of the correspondences that it could theoretically know from its ontology to others. Finally we consider that the histories of received queries constantly change over time: scoring function values vary continually. It is considered as a critical situation.

6.2   Metrics

In order to measure the efficiency of the CORDIS protocol we consider the richness measure $\mathcal{H}_{Rich}$ (see equation (3)) and the measures that allows to capture the heterogeneity related to the disparity between peers: $\mathcal{H}_{Disp}$ and $\mathcal{H}_{DapAvg}$ (see equations (4) and (8)). Remember that richness ($\mathcal{H}_{Rich}$) characterizes the diversity of a population and is linked to the number of ontologies used in the system. The "structure unaware, disparity aware" measure $\mathcal{H}_{Disp}$ reflects how much the peers are globally disparate from each others. The "structure aware, disparity aware" measure $\mathcal{H}_{DapAvg}$ reflects how much the peers are surrounded by disparate peer.

6.3   Experiment 1

In this experiment we study the behaviour of CORDIS in different situations of semantic diversity (which can be measured with $\mathcal{H}_{Rich}$). We consider unstructured P2P systems of 500 peers, and we vary the number of used ontologies: 23, 46 or 93. As a consequence, $\mathcal{H}_{Rich}$ equals 0.045, 0.09 or 0.18. These values seem low but actually they represent relatively critical situations. For instance, when 93 ontologies are used, it means that each ontology is shared by only 5.3 peers (in average). We do not limit the size of the repository (peers can store all the o_p-correspondences), and we limit the size of the cache to 200 entries. At each cycle, peers exchange messages, for which the size is limited to 75 entries: 50 come from the repository, and 25 from the cache (*i.e.* we set $\pi = 2/3$). Table 3 summarizes this set of parameters.

Table 3. Simulation parameters

Parameter	Value(s)
# of peers	500
# of ontologies	93, 46, 23
Size of the repository	∞
Size of the cache	200
Size of messages	75

Figures 4 and 5 present the evolution of $\mathcal{H}_{Disp}$ and $\mathcal{H}_{DapAvg}$ w.r.t. the number of cycles produced by the CORDIS protocol. Horizontal lines correspond to the optimal theoretical values that would be obtained if all peers knew all the possible correspondences involving their ontologies. It is important to point out that the results presented in this section are very related to the dataset used. As there are very few correspondences between ontologies, the impact of CORDIS on $\mathcal{H}_{Disp}$ and $\mathcal{H}_{DapAvg}$ may seem modest. The evolution of $\mathcal{H}_{Disp}$ and $\mathcal{H}_{DapAvg}$ are similar. This is due to the fact that systems are randomly organized: they are not organized according to ontologies used by peers. The figures show that when heterogeneity $\mathcal{H}_{Rich}$ increases, the CORDIS protocol becomes less efficient ($\mathcal{H}_{Disp}$ and $\mathcal{H}_{DapAvg}$ are less reduced). Still CORDIS allows to reduce heterogeneity ($\mathcal{H}_{Disp}$ and $\mathcal{H}_{DapAvg}$) from 0.94 to 0.78 when $\mathcal{H}_{Rich}$ is high (*i.e.* equal to 0.18). In all cases of semantic richness, heterogeneity converges and tends to the optimal theorical value after respectively 250, 300 and 350 cycles for the heterogeneity $\mathcal{H}_{Disp}$[4].

---

[4] Values are different for $\mathcal{H}_{DapAvg}$ but comments are similar.

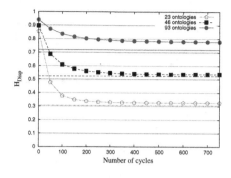 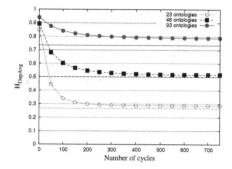

Fig. 4. Evolution of $\mathcal{H}_{Disp}$ in systems of 500 peers

Fig. 5. Evolution of $\mathcal{H}_{DapAvg}$ in systems of 500 peers

6.4   Experiment 2

In this experiment we study the behaviour of CORDIS according to the size of the system. We consider three situations, in which 93 ontologies are used. The number of peers are respectively 250, 500 and 1,000. As the number of ontologies is fixed, the semantic diversity ($\mathcal{H}_{Rich}$) equals 0.36, 0.18 and 0.09. The size of the repository, the cache, and the messages are set as in the previous experiments (see Table 3).

Figures 6 and 7 respectively present the evolution of $\mathcal{H}_{Disp}$ and $\mathcal{H}_{DapAvg}$ w.r.t. the number of cycles. Theoretical optimal values are not explicitly shown on figures but like in the previous experiment it is clear that heterogeneity cannot be reduced under a certain value related to the dataset we consider. In both cases ($\mathcal{H}_{Disp}$ and $\mathcal{H}_{DapAvg}$) we can see that heterogeneity is more reduced when the number of peers is important. This clearly shows that if the number of ontologies on use is fixed, the fact that peers are numerous favors the reduction of heterogeneity. This can be explained by the fact that correspondences are replicated in the system. Each correspondence is stored by a more important number of peers. Then it is easily "accessible" by peers that need it. There is no problem of scalability when using the CORDIS protocol to reduce heterogeneity. The convergence time is more important when there are 1,000 peers in the system than when there are just 250 peers. More generally, the bigger the system, the longer the convergence time. By cons, we can see that after 100 cycles, heterogeneity values obtained when there are 1,000 peers are lower than values obtained after converge when there are only 250. The conclusion is that increasing the number of peers can further reduce the heterogeneity even if convergence is longer.

6.5   Experiment 3

In this experiments we study the behaviour of CORDIS according to the dynamicity of the system. The dynamicity is a key-property of P2P systems. It comes from the fact that peers can join or leave the system at any time. We consider

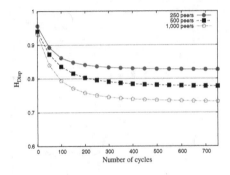 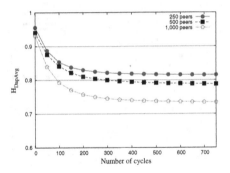

Fig. 6. Evolution of $\mathcal{H}_{Disp}$ in which 93 ontologies are used

Fig. 7. Evolution of $\mathcal{H}_{DapAvg}$ in systems in which 93 ontologies are used

P2P systems of 500 peers using 93 different ontologies. As a consequence, the semantic diversity $\mathcal{H}_{Rich}$ equals 0.09. The degree of connectivity is set to 4. In order to simulate the dynamicity (also known as the churn effect), we make some peers join and leave the system at each cycle. As we want the number of peers to be constant during the simulation, we consider that the number of peers joining the system at each cycle equals the number of peers that are leaving. We study the performance of CORDIS when the session duration average varies between 1 minute and 15 minutes. The session duration of a peer corresponds to the time this latter remains in the system. Peers joining the system use ontologies that are potentially already used in the system. We consider that the length of a cycle is five seconds (12 cycles per minute). Our experiment consists in observing a system during 750 cycles (62.5 minutes). We considered the configurations presented in Table 4. If the session duration is set to m minutes (*i.e.* ($m \times 12$) cycles), it means that $\frac{|\mathcal{P}|}{m \times 12}$ peers leave at each cycle. In this situation the churn rate is ($\frac{100}{m \times 12}$)%. In this context a churn rate greater than 1% is a critical and unlikely case because it means that peers remain only eight minutes in the system in average (this time does not seem sufficient to share or download data).

Table 4. Configurations considered to study CORDIS with churn

Session duration (minute)	(cycle)	Churn rate (%)
1	12	8.33
5	60	1.67
15	180	0.54
∞	∞	0 (no churn)

Figures 8 and 9 present heterogeneity values ($\mathcal{H}_{Disp}$ and $\mathcal{H}_{DapAvg}$) w.r.t. the dynamicity of the system. The case where the churn rate equals 0% corresponds

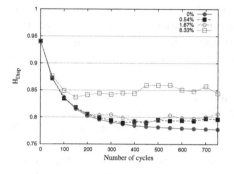

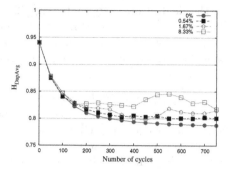

Fig. 8. Evolution of $\mathcal{H}_{Disp}$ in systems of 500 peers in which 93 ontologies are used

Fig. 9. Evolution of $\mathcal{H}_{DapAvg}$ in systems of 500 peers in which 93 ontologies are used

to the reference (no churn). It is important to notice that dynamicity cannot help in reducing more heterogeneity because some correspondences are deleted when peers leave the system. Results show that heterogeneity reduction is almost not impacted when sessions last 5 or 15 minutes. When sessions last only one minute, heterogeneity is less reduced. It is due to the fact that peers do not have enough time to learn interesting correspondences. Heterogeneity $\mathcal{H}_{Disp}$ is reduced from 0.94 to 0.85, and $\mathcal{H}_{DapAvg}$ is reduced from 0.94 to 0.83. In this case heterogeneity is not stable because peers with a lot of correspondences may have left the system. This globally penalized the system heterogeneity. As a conclusion we can say that CORDIS is suitable for dynamic systems: even if the session duration is very low (a single minute), heterogeneity is reduced.

7   Related Work

The work presented in this article is a revised and extended version of [12]. It presents the model and the CorDis algorithm in a more detailed way and it in addition provides a typology of semantic heterogeneity measures with definitions of each type. Section 5 is new as the first version did not consider the possible inconsistency of peers' knowledge bases. Finally, the experimentations still study the evolution of different facets of heterogeneity in function of several factors, but we switched from the OntoFarm data set to a bigger set of 93 ontologies from the BioPortal, that are actively used in the biomedical domain.

7.1   Results Used in Our Work

Our work builds on previous results obtained in several fields which we recall here.

Distance between ontologies. Our measures of semantic heterogeneity assume the existence of a disparity measure between peers. Distance measures proposed in the field of ontology matching [22,34,18,19] can be adapted and considered as disparity measures. However, in general a disparity measure between two peers cannot be reduced to a distance between their ontologies as other elements might be considered like for example the similarity between concepts of a peer's ontology [13].

Gossiping protocols. In P2P systems, gossiping algorithms are simple, robust and flexible solutions to ensure several tasks like information dissemination, peer sampling or overlays maintenance [31,30,29]. Whereas a general framework is proposed, one has to adapt it for its own application. This is what we have done with CORDIS, adapting dissemination to correspondence sharing with a mechanism that considers the history of queries to rank correspondences and rule out the less interesting ones when storage space lacks. This type of dissemination algorithm has good properties but may be improved to avoid redundancy [31]. This is our of the scope of this paper, as we focus on the effect of disseminating correspondences to impact the heterogeneity measures.

Dealing with inconsistencies. In the context of Description Logics, we use a general definition of inconsistency and refer to [7] for more definitions and results. Detecting inconsistency, repairing an inconsistent set, extracting modules that keep consistent w.r.t. the initial ontology is a research area in its own [39,40,41,43,28,32,24]. We do not provide new results in this field. We just pick up the elements that enable us to define possible peers policies when they are confronted to an inconsistency and we reconsider defining semantic heterogeneity in such a case. To our knowledge, this has not been studied in such a context.

7.2   Comparison with Other Works

We underline the differences with several works and systems. The ultimate goal of all these works, including our, is to provide a good interoperability between the peers, *i.e.* to ensure they are able to understand each other's queries and results. However, the approaches differ on the hypotheses and context. A first way to ensure good interoperability is to create a global ontology that serves as an intermediary between all peers of the system [20]. Although it is valuable in some cases, this approach may have problems to scale up.

Considering that each peer may choose the ontology that best fits its needs is closer to the hypothesis made for *Peer Data Management Systems* (PDMS) [26,16,11]. It is moreover assumed that neighbor peers know mappings between their schemas, if such mappings can be established. However, unlike our work, the transitivity of schema mappings is exploited to retrieve the entire network, through successive translations done by the receiving peers. In this context, notice that the term *semantic gossiping* has been used to refer to the action of "propagating queries toward nodes for which no direct translation link exists" [1,2]. In our work, semantic gossiping disseminates the correspondences in the system for the peers to share their knowledge and it is totally independent of the query evaluation. This corresponds to our view of a layered system, where correspondences sharing and query evaluation lie in different layers, the former being the only focus of this paper. This motivates the introduction of heterogeneity measures that are able to characterize the peers' semantic network and the effects of the gossiping algorithm. The independency between the two layers also enables to think of query evaluation protocols that let the peers master their own translation choices.

Works like [26,11] mainly focus on expressing schema mappings and translation of queries expressed in some language like XQuery, using the mappings known between neighbor peers. Once more, translation is not our focus here. However, CorDis provides a means to share correspondences that could be used for example in [11] to avoid a low coverage of a query by a chain of mappings. Indeed, assuming a query can be translated from one to another all along a path requires at least a strong connectivity condition. This is highlighted in [15] where the authors define a necessary condition for interoperability. This condition has the same kind of goal as our heterogeneity measures as it tries to characterize the semantic network but is quite different by nature. However, we could add a mechanism in a system based on CorDis in order to check it.

In [17] authors propose a system ensuring interoperability by offering several functionalities to automatically organize the network of mappings at a mediation layer. In particular it ensures that the network of schemas and mappings is strongly connected. If the number of mappings in not sufficient to ensure the connectivity, mappings are automatically created (the way it is done is not detailed). This work can be considered as complementary to ours. Pires *et al.* [38] present a semantic matcher which identifies correspondences between ontologies used in a PDMS. This method could be used in our context to discover correspondences, *i.e.* to initialize peers' alignments or to enrich them.

Finally, it is possible to group peers into Semantic Overlay Networks (SONs) [14] according to their disparity. This makes closer the peers that are similar in some way. As a consequence, semantic heterogeneity based on disparity is lowered and interoperability is increased. The topology adaptation may be guided through a predefined hierarchy [14] or more dynamic [36]. These type of topology adaptation protocols are perfect candidates to run in complement to CorDis in order to implement the low layer of a system that would manage the decrease of heterogeneity. Then a higher layer would implement the query evaluation protocol that enables interoperability.

8   Conclusion

In this work, we presented a new approach to address the problem of semantic heterogeneity of unstructured P2P systems. This approach consists in considering separately the problem (that is to say the heterogeneity) and the consequence of the problem (the difficulty to interoperate). We focused on the semantic heterogeneity, which is directly related to the fact that peers use different ontologies to represent their resources. We first proposed a set of measures to characterize different facets of heterogeneity. The idea of these measures is to capture the situation of heterogeneity of a system, and to allow measuring the contribution of an algorithm that aims to decrease it. Second we proposed a protocol called CORDIS that allows to reduce some facets of semantic heterogeneity. Indeed this gossip-based protocol reduces the heterogeneity related to disparities between peers. It makes peers share correspondences known by some peers but ignored by others. Because sharing knowledge might lead to inconsistencies, we proposed different peer's policy to deal with inconsistencies. We also reconsidered the definitions of measures of heterogeneity in the case where there are inconsistencies. The experiments represented a great deal of work, as they considered 93 ontologies that required some preliminary manipulations and different evaluation contexts. Although no one can yet say what a "real" semantic P2P system looks like, the experiments not only had the advantage to consider rather numerous ontologies but also ontologies that are actively used in the biomedical domain. We showed that the CORDIS protocol is effective in different situations of semantic diversity and that it scales and manages the dynamicity of the system.

As future work, we first plan to add a mechanism to CORDIS to trigger alignment processes between two ontologies when it is needed. In our assumptions, we have considered that alignments were initially calculated between neighbors. In this situation two peers that exchange queries and that are not directly neighbors may not have correspondences to translate queries (and may not be in a position to learn correspondences) . In this case, it would be wise to encourage these two peers to align their ontologies. We also plan to define additional algorithms to reduce other facets of semantic heterogeneity. In particular we will work to define a protocol to organize P2P systems according to peers' semantic knowledge. This should help to reduce the heterogeneity captured by topology-aware measures. This new protocol would be perfectly complementary to CORDIS because one reduces the heterogeneity related to disparities between peers, and the other reduces the one related to the system topology. A clever combination of the two algorithms should even improve the results.

These first steps fit our long-term goal of designing and implementing resource sharing P2P systems with a modular two layered architecture. A low layer would manage semantic heterogeneity and ensure its decrease. It could use several complementary protocols (CORDIS and other ones). An upper layer would focus on ensuring interoperability through smart query evaluation algorithms. In practice, this would enable the participants to annotate their resources using the ontology they prefer, while still being able to automatically both answer queries and retrieve resources annotated with other ontologies.

References

1. Aberer, K., Cudré-Mauroux, P., Hauswirth, M.: The chatty web: emergent semantics through gossiping. In: WWW, pp. 197–206 (2003)
2. Aberer, K., Cudré-Mauroux, P., Hauswirth, M., Van Pelt, T.: GridVine: Building Internet-Scale Semantic Overlay Networks. In: McIlraith, S.A., Plexousakis, D., van Harmelen, F. (eds.) ISWC 2004. LNCS, vol. 3298, pp. 107–121. Springer, Heidelberg (2004)

3. Akbarinia, R., Pacitti, E., Valduriez, P.: Reducing network traffic in unstructured P2P systems using top-k queries. Distributed and Parallel Databases 19(2-3), 67–86 (2006)

4. Akbarinia, R., Pacitti, E., Valduriez, P.: Query processing in P2P systems. Research report, INRIA (2007)

5. Antezana, E., Egaña, M., De Baets, B., Kuiper, M., Mironov, V.: ONTO-PERL: An API for supporting the development and analysis of bio-ontologies. Bioinformatics 24(6), 885–887 (2008)

6. Antoniou, G., van Harmelen, F.: Web ontology language: OWL. In: Staab, S., Studer, R. (eds.) Handbook on Ontologies, International Handbooks on Information Systems, 2nd edn., pp. 91–110. Springer, Heidelberg (2009)

7. Baader, F., Calvanese, D., McGuinness, D.L., Nardi, D., Patel-Schneider, P.F. (eds.): The Description Logic Handbook: Theory, Implementation, and Applications. Cambridge University Press (2003)

8. Baeza-Yates, R., Ribeiro-Neto, B.: Modern Information Retrieval. Addison Wesley (May 1999)

9. Bechhofer, S., Volz, R., Lord, P.: Cooking the Semantic Web with the OWL API. In: Fensel, D., Sycara, K., Mylopoulos, J. (eds.) ISWC 2003. LNCS, vol. 2870, pp. 659–675. Springer, Heidelberg (2003)

10. Bellahsene, Z., Bonifati, A., Rahm, E. (eds.): Schema Matching and Mapping. Springer (2011)

11. Bonifati, A., Chang, E., Ho, T., Lakshmanan, L.V.S., Pottinger, R., Chung, Y.: Schema mapping and query translation in heterogeneous P2P XML databases. The VLDB Journal 19, 231–256 (2010)

12. Cerqueus, T., Cazalens, S., Lamarre, P.: Gossiping Correspondences to Reduce Semantic Heterogeneity of Unstructured P2P Systems. In: Hameurlain, A., Tjoa, A.M. (eds.) Globe 2011. LNCS, vol. 6864, pp. 37–48. Springer, Heidelberg (2011)

13. Cerqueus, T., Cazalens, S., Lamarre, P.: Semantic heterogeneity measures of unstructured P2P systems. In: 10th IEEE/WIC/ACM International Conference on Web Intelligence, pp. 223–226 (2011)

14. Crespo, A., Garcia-Molina, H.: Semantic Overlay Networks for P2P Systems. In: Moro, G., Bergamaschi, S., Aberer, K. (eds.) AP2PC 2004. LNCS (LNAI), vol. 3601, pp. 1–13. Springer, Heidelberg (2005)

15. Cudré-Mauroux, P., Aberer, K.: A Necessary Condition for Semantic Interoperability in the Large. In: Meersman, R. (ed.) CoopIS/DOA/ODBASE 2004. LNCS, vol. 3291, pp. 859–872. Springer, Heidelberg (2004)

16. Cudré-Mauroux, P., Agarwal, S., Aberer, K.: GridVine: An infrastructure for peer information management. IEEE Internet Computing 11(5), 36–44 (2007)

17. Cudré-Mauroux, P., Agarwal, S., Budura, A., Haghani, P., Aberer, K.: Self-organizing schema mappings in the GridVine peer data management system. In: 33rd International Conference on Very Large Data Bases, pp. 1334–1337 (2007)

18. David, J., Euzenat, J.: Comparison between Ontology Distances (Preliminary Results). In: Sheth, A.P., Staab, S., Dean, M., Paolucci, M., Maynard, D., Finin, T., Thirunarayan, K. (eds.) ISWC 2008. LNCS, vol. 5318, pp. 245–260. Springer, Heidelberg (2008)

19. David, J., Euzenat, J., Šváb-Zamazal, O.: Ontology Similarity in the Alignment Space. In: Patel-Schneider, P.F., Pan, Y., Hitzler, P., Mika, P., Zhang, L., Pan, J.Z., Horrocks, I., Glimm, B. (eds.) ISWC 2010, Part I. LNCS, vol. 6496, pp. 129–144. Springer, Heidelberg (2010)

20. De Souza, H.C., De C. Moura, A.M., Cavalcanti, M.C.: Integrating ontologies based on P2P mappings. IEEE Transactions on Systems, Man, and Cybernetics 40, 1071–1082 (2010)
21. Eberspächer, J., Schollmeier, R.: First and Second Generation of Peer-to-Peer Systems. In: Steinmetz, R., Wehrle, K. (eds.) P2P Systems and Applications. LNCS, vol. 3485, pp. 35–56. Springer, Heidelberg (2005)
22. Euzenat, J., Shvaiko, P.: Ontology matching. Springer (2007)
23. Noy, N.F., Shah, N.H., Whetzel, P.L., Dai, B., Dorf, M., Griffith, N., Jonquet, C., Rubin, D.L., Storey, M.-A.D., Chute, C.G., Musen, M.A.: Bioportal: ontologies and integrated data resources at the click of a mouse. Nucleic Acids Research 37(Web-Server-Issue), 170–173 (2009)
24. Goasdoué, F., Rousset, M.-C.: Robust module-based data management. IEEE Transaction on Knowledge and Data Engineering (2012)
25. Gruber, T.R.: A translation approach to portable ontology specifications. Knowledge Acquisition 5, 199–220 (1993)
26. Halevy, A.Y., Ives, Z., Mork, P., Tatarinov, I.: Piazza: data management infrastructure for semantic web applications. In: 12th International World Wide Web Conference, pp. 556–567 (2003)
27. Hartung, M., Terwilliger, J.F., Rahm, E.: Recent advances in schema and ontology evolution. In: Schema Matching and Mapping, pp. 149–190. Springer (2011)
28. Hunter, A., Sébastien, K.: Measuring inconsistency through minimal inconsistent sets. In: Proceedings of the Eleventh International Conference on Principles of Knowledge Representation and Reasoning, pp. 358–366 (2008)
29. Jelasity, M., Babaoglu, O.: T-man: Fast gossip-based constructions of large-scale overlay topologies. Technical Report UBLCS-2004-7 (2004)
30. Jelasity, M., Guerraoui, R., Kermarrec, A.-M., van Steen, M.: The Peer Sampling Service: Experimental Evaluation of Unstructured Gossip-Based Implementations. In: Jacobsen, H.-A. (ed.) Middleware 2004. LNCS, vol. 3231, pp. 79–98. Springer, Heidelberg (2004)
31. Kermarrec, A.-M., van Steen, M.: Gossiping in distributed systems. Operating Systems Review 41(5), 2–7 (2007)
32. Ma, Y., Qi, G., Hitzler, P.: Computing inconsistency measure based on paraconsistent semantics. Journal of Logic and Computation 21(6), 1257–1281 (2011)
33. Ma, Y., Qi, G., Hitzler, P., Lin, Z.: Measuring Inconsistency for Description Logics Based on Paraconsistent Semantics. In: Mellouli, K. (ed.) ECSQARU 2007. LNCS (LNAI), vol. 4724, pp. 30–41. Springer, Heidelberg (2007)
34. Maedche, A., Staab, S.: Measuring Similarity between Ontologies. In: Gómez-Pérez, A., Benjamins, V.R. (eds.) EKAW 2002. LNCS (LNAI), vol. 2473, pp. 251–263. Springer, Heidelberg (2002)
35. Montresor, A., Jelasity, M.: Peersim: A scalable P2P simulator. In: 9th IEEE International Conference on Peer-to-Peer Computing, pp. 99–100 (2009), http://peersim.sf.net
36. Penzo, W., Lodi, S., Mandreoli, F., Martoglia, R., Sassatelli, S.: Semantic peer, here are the neighbors you want! In: 11th International Conference on Extending Database Technology, pp. 26–37 (2008)
37. Peroni, S., Motta, E., d'Aquin, M.: Identifying Key Concepts in an Ontology, through the Integration of Cognitive Principles with Statistical and Topological Measures. In: Domingue, J., Anutariya, C. (eds.) ASWC 2008. LNCS, vol. 5367, pp. 242–256. Springer, Heidelberg (2008)

38. Pires, C.E., Souza, D., Pachêco, T., Salgado, A.C.: A Semantic-Based Ontology Matching Process for PDMS. In: Hameurlain, A., Tjoa, A.M. (eds.) Globe 2009. LNCS, vol. 5697, pp. 124–135. Springer, Heidelberg (2009)
39. Reiter, R.: A theory of diagnosis from first principles. Artificial Intelligence 32(1), 57–95 (1987)
40. Schlobach, S., Huang, Z., Cornet, R., Van Harmelen, F.: Debugging incoherent terminologies. Journal of Automated Reasoning 39(3), 317–349 (2007)
41. Sirin, E., Parsia, B., Grau, B.C., Kalyanpur, A., Katz, Y.: Pellet: A practical owl-dl reasoner. Journal of Web Semantics 5(2), 51–53 (2007)
42. Steinmetz, R., Wehrle, K. (eds.): P2P Systems and Applications. LNCS, vol. 3485. Springer, Heidelberg (2005)
43. Tsarkov, D., Horrocks, I.: FaCT++ Description Logic Reasoner: System Description. In: Furbach, U., Shankar, N. (eds.) IJCAR 2006. LNCS (LNAI), vol. 4130, pp. 292–297. Springer, Heidelberg (2006)

Towards a Scalable Semantic Provenance Management System

Mohamed Amin Sakka[1,2] and Bruno Defude[2]

[1] Novapost, Novapost R&D, 32, Rue de Paradis 75010 Paris-France
amin.sakka@novapost.fr
[2] TELECOM SudParis, CNRS UMR Samovar,
9, Rue Charles Fourier 91011 Evry cedex-France
{mohamed_amin.sakka,bruno.defude}@it-sudparis.eu

Abstract. Provenance is a key metadata for assessing electronic documents trustworthiness. It gives an indicator on the reliability and the quality of the document content. Most of the applications exchanging and processing documents on the web or in the cloud become provenance aware and provide heterogeneous, decentralized and not interoperable provenance data. Most of provenance management systems are either dedicated to a specific application (workflow, database) or a specific data type. Those systems were not conceived to support provenance over distributed and heterogeneous sources. This implies that end-users are faced with different provenance models and different query languages. For these reasons, modeling, collecting and querying provenance across heterogeneous distributed sources is still considered as a challenging task.

This work presents a new provenance management system (PMS) based on semantic web technologies. It allows to import provenance sources, to enrich them semantically to obtain high level representation of provenance. It supports semantic correlation between different provenance sources and allows the use of a high level semantic query language. In the context of cloud infrastructure where most of applications will be deployed in a near future, scalability is a major issue for provenance management systems. We described here an implementation of our PMS based on an NoSQL database management system coupled with the map-reduce parallel model and show that it scales linearly depending on the size of the processed logs.

1 Introduction and Motivations

1.1 Introduction

The provenance of a piece of data is the process that lead to that piece of data. It is metadata recording the ultimate derivation and passage of an item through its various owners. The description of such a derivation may take different forms or may emphasize different properties. Provenance can be used in different application areas and for different purposes like reliability, quality, re-usability, justification, audit, migration. Provenance provides an evidence of authenticity,

A. Hameurlain et al. (Eds.): TLDKS VII, LNCS 7720, pp. 96–127, 2012.

integrity and information quality [1]. Some works [2] use it to improve business processes (monitoring and optimization aspects of BPM) and to ensure the traceability of end-to-end operations. Other works have develop a new generation of database [3] and file systems [4] taking into account the provenance dimension especially for regulatory compliance.

With the maturation of service oriented technologies and cloud computing, more and more data are exchanged electronically and dematerialization becomes one of the key concepts to cost reduction and efficiency improvement. Different services used for creating, processing, exchanging and archiving documents become available and accessible. However, in a legal archiving context, in order to contribute to confidentiality and privacy, the provenance of the documents should be recorded and made accessible when needed (audit, litigation about the authenticity, SLA[1] ...). Thus, many research works give extreme importance to data provenance and document life cycle management [5,6].

In general, provenance data is distributed across different log files. These logs are structured using different formats (database, text, XML ...), are queried using a suitable query language (SQL, XQuery ...), are semantically heterogeneous and can be distributed across servers. For example, an application can produce provenance data in a generic log file like a http log (generated by a http server) and in a dedicated (application-oriented) one (generated by application). These provenance sources need to be semantically correlated in order to produce an integrated view of provenance. Examples of such correlations are correspondences between different items. To support that, existing provenance models are based on one of two approaches: either put provenance and business specificities in a unique and semantically rich model or propose a basic model that can be semantically enriched with different techniques. Moreover, if we consider complex business processes or complex document processing workflows, they can be considered as orchestrations of applications thus increasing the number of provenance sources, their heterogeneity and distribution, as well as, the need for semantic correlations.

For these reasons, we need provenance management systems (or PMS) addressing the following points:

- support of syntactic and semantic heterogeneity of provenance sources.
- support of rich domain models allowing to construct high level representation of provenance.
- support of semantic correlation between different domain models.
- support of high level semantic query languages.
- scalability.

The main contributions of this work are: (1) the proposition of a provenance framework for log integration and enrichment based on semantic web standards and tools, (2) a flexible model to describe provenance and business specificities based on domain models and instantiated domain models, (3) a fusion process allowing to integrate multiple and heterogeneous provenance sources into an

---

[1] SLA: Service Level Agreement.

integrated domain model, (4) the use of the SPARQL standard to query semantically enriched logs and (5) two implementations of a PMS, the first one based
on an RDF store offering a high level of expressiveness but lacking scalability
and the second one, based on a NoSQL DBMS scaling up linearly with a limited
decrease in expressiveness.

The rest of the paper is organized as follows. In section 2 we start by detailing
the components of the proposed framework based on semantic web technologies.
Then, we illustrate how the distributed aspect of provenance and interoperability
issues will be managed within our framework. Section 3 details an industrial
use case coming from the legal archiving context (Novapost) where provenance
management is extremely important. Then, it presents the application of the
proposed approach to that use case. Section 4 is composed by two parts. The
first one presents the logical architecture of the provenance management system
whereas the second one presents the implementation of the architecture using
two different technologies and a set of experimentations. In these experiments,
we compare provenance data load time and the average query execution time for
different dataset size. Finally, we discuss related work in section 5 and conclude
and describe future work in section 6.

2   A Provenance Management Framework Based on Semantic Web Technologies

2.1   Overview

In this part, we describe how our proposal fulfill the aforementioned objectives.
We aim to define a provenance framework offering principally two degrees of
liberty: (i) an interpretation of heterogeneous provenance data with a common
domain model, and (ii) different levels of interpretation with different domain
models of the same provenance data.

To address these issues and facilitate provenance management, we think that
a semantic provenance framework is a suitable solution. In fact, combining different data sources has never been easy but the semantic web enables data to be
joined more easily across boundaries. Further, semantic web technologies support complex domain concept modeling with OWL[2] what permits to describe
any business specificity. Using RDF to share and store provenance enables its interoperability and permits to query it based on SPARQL [3] standard. Within our
framework, the provenance of any document is extracted from log files (see Figure 1 part A) and is represented using a semantic model. The basis of this model
is the minimal domain model or MDM (see Figure 1 part B). It contains only
provenance informations without any reference to a particular domain (what is
the action performed, on which document, by whom and when). It is based on
the Open Provenance Model (OPM)[4] [7]. This model ensures the genericity of

---

[2] Web Ontology Language: http://www.w3.org/TR/owl-features
[3] http://www.w3.org/TR/rdf-sparql-query
[4] Open Provenance Model: http://openprovenance.org/

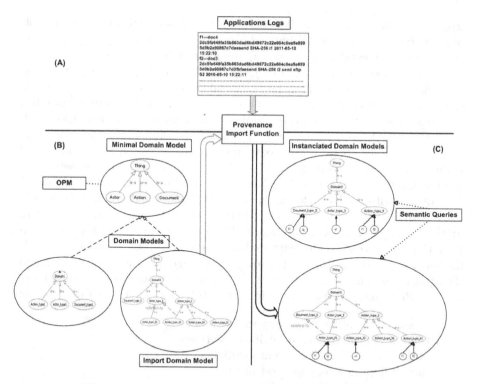

Fig. 1. Components of the semantic provenance framework

our approach and the support of basic semantic heterogeneity between logs. The MDM is specialized to define different domain models and describe their domain constraints (see Figure 1 part B). These models allow to support semantic heterogeneity and to give different views of the same log.

Most of real applications will not use these models as their native log formats. Consequently, provenance providers should define an import function taking as input a provenance source and the corresponding domain model (called import domain model) and producing an instantiated domain model (see Figure 1 part C). This model will be queried by end-users with a semantic query language like SPARQL.

Between the instantiated import domain model and the instantiated minimal domain model, a lattice of instantiated domain models can be created by means of a *generalization* relationship. This relationship permits to automatically generate a new, more abstract instantiated domain model that can be also queried using SPARQL.

2.2 OPM as a Generic Model for Provenance

OPM [7] is a multi-layer provenance model which describes the data and activity dependencies of all the participants. In essence, it consists of a directed graph expressing such dependencies. It defines a set of nodes: *artifact* (an immutable piece of state), *process* (an activity or a set of activities) or *agent* (contextual entity acting as a catalyst of a process) and a set of dependencies between them such as *(process) used (agent), (artifact) wasGeneratedBy (process), (process) wasControlledBy (agent)*, etc. . . These causal dependencies are represented as edges between the entities from the effect to the cause and they are composable. OPM can be semantically enriched using accounts, profiles and annotations. We admit that OPM has a strong focus on scientific workflows. Its applicability in scenarios involving users, where computations take place on the desktop and in the cloud, where various forms of artifacts are manipulated (datasets, files, documents, etc. . .) has not been demonstrated yet. Enriched OPM graphs mixes the semantic with the basic domain model what limits the flexibility of interpretation with different domain models. At this level, we propose to use a subset of OPM as a basis to represent provenance. We omit accounts (and thus the notion of overlapping accounts, alternate and refinement) and annotations. We omit also the use of two edge types: (1) process was triggered by another process and (2) artifact was generated by another artifact. Formally, we define a provenance graph by ProvenanceGraph(ProcessId, ArtifactId, AgentId, Value, Edge) where:

- ProcessId: a primitive set of process identifiers.
- ArtifactId: a primitive set of artifact identifiers.
- AgentId: a primitive set of agent identifiers.
- Value: an application-specific set of values.
- Edge $\subseteq$ Used $\cup$ WasGeneratedBy.
- Used: ProcessId $\times$ Role $\times$ ArtifactId.
- WasGeneratedBy: ArtifactId $\times$ Role $\times$ ProcessId.

Based on this formalization, an instance of our provenance graph (which is already an OPM compliant graph) is therefore a tuple of 7 elements defined by
$$pg = <ProcessId, ArtifactId, AgentId, Value, Role, Edge, valueOf>$$
where:

- Role is a primitive set of roles.
- valueOf is of type ValueOf where ValueOf: Node $\rightarrow$ $\mathbb{P}$(Value).
- Node = ProcessId $\cup$ ArtifactId.

This model is the basis of our definition of domain models and acts as a minimal common representation among all logs, ensuring interoperability.

2.3   Domain Models

Within our framework, domain specificities and constraints are encapsulated in an ontology. For standardization reasons, we use OWL as a formalism for domain model ontology definition. Our restricted OPM is used to define the basis of all domain models and we propose to encode it as a simple OWL ontology called Minimal Domain Model (MDM for short). Any domain model should be defined by specializing it.

Definition 1: Minimal Domain Model (MDM): The minimal domain model is described by an ontology expressed in OWL and is defined by:

- a set of three disjoint classes BC: *Document, Action, Actor.*
- a set of data types D: *documentId, actionId, actorId, applicationId, timestamp.*
- a set of data type properties TP: *hasDocumentId, hasActionId, hasActorId, hasApplicationId , hasTimeStamp, precededBy.* Each data type property has a domain in BC and a range in D as follows:
  - hasDocumentId: Document $\rightarrow$ documentId.
  - hasActionId: Action $\rightarrow$ actionId.
  - hasActorId: Actor $\rightarrow$ actorId.
  - hasApplicationId: BC $\rightarrow$ applicationId.
  - hasTimestamp: Action $\rightarrow$ timestamp.
  - precededBy: Action $\rightarrow$ Action.

Definition 2: Domain model (DM)

A domain model is a specialization of the MDM. It allows to describe the business specificities of a specific domain that are not covered by the MDM. This specialization consists in the addition of new concepts extending MDM concepts, new data types or new relationships. Thus, it becomes possible to describe specific domain documents, actions, actors, their dependencies and properties. A specialized domain model DM_i, specialization of the MDM (we can also say that MDM is a generalization of DM_i), $DM_i \leq_{spec}$ MDM is defined as follows:

- a set of sub-classes SC extending MDM basic classes BC using the inheritance function *is-a* as defined in OWL.
- a set of data types D.
- a set of data type properties TP. Each data type property has a domain in SC and a range in D.
- a set of object properties OP. Each object property has its domain and range in SC.
- a set of subclass relationships over (SC×SC).

This definition is valid for other DMs than MDM.

2.4   Instantiated Domain Models

An instantiated domain model IDM is a DM populated with provenance data. It contains the trace of the different actions, their input, output and the actors performing them. It also contains different data type values. These instances can be imported from a unique or multiple provenance sources using an import function defined by a domain specialist (e.g application administrator). This function is essential in our framework. Its principal role is to instantiate the import domain model by generating provenance instances. To enhance syntactic interoperability between heterogeneous provenance sources, we choose RDF for creating and storing provenance instances.

Definition 3: Instantiated Domain Model (IDM)

An IDM is associated to a specific domain model DM and a unique or multiple provenance sources. It is generated by an import function and is an OWL ontology which is an extension of the DM that includes instances of documents, actors and actions linked to the domain concepts with *owl:InstanceOf* function. Figure 2 gives the example of a specific domain model (from Novapost) and a sample associated IDM.

To interpret the imported provenance (IDM_{import}) using another domain model (DM_{target}), we need to create a new IDM_{target} corresponding to that DM. This latter can be obtained by applying generalization rules permitting to generate it automatically from IDM_{import}. This generalization can be viewed as a new classification of IDM_{import} consisting on projecting some instances to an upper level and deleting some others.

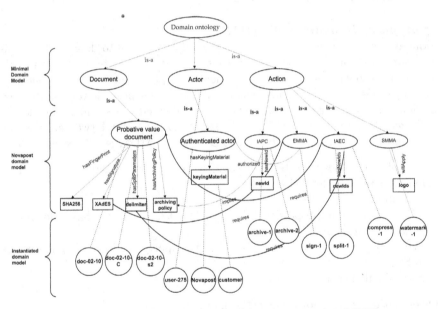

Fig. 2. The Novapost domain model and an associated IDM

Definition 4: Generalization of an IDM

We define the generalization relationship for an instantiated domain model $\leq$: IDM $\rightarrow$ IDM and we notice $IDM_j \leq IDM_i$ (IDM_i is a generalization of IDM_j or IDM_j is a specialization of IDM_i) if and only if:

- For each class of IDM_j: all instances of IDM_j becomes instances of the parent class defined in IDM_i (extra data types and data types properties that are not defined in the target domain model DM_i for that class will be deleted).

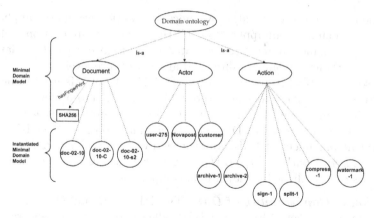

Fig. 3. An example of IDM$_{import}$ generalization to the MDM as target model

– For each relationship between DM_j classes: all the relationships between
  two concepts that does not exist between the parent concepts in the target
  domain model DM_i will be deleted.

An example of IDM *generalization* is given in Figure 3. The *generalization* relationship on IDMs allows to interpret the same provenance source with different
domain models but using a single import function, based on the most specialized
domain model. Of course, it is also possible to define different import functions
for the same provenance source but this requires much more effort from administrators.

Definition 5: Lattice of IDMs: The generalization relationship defines a
lattice of IDMs with the import IDM as the lower bound and the instantiated
MDM as the upper bound.

The introduction of semantic DMs and IDMs allows to address semantic heterogeneity between provenance sources, all sources have in common at least the
MDM that can be queried using a powerful query language such as SPARQL.
The *specialization* relationship on DMs and *generalization* relationship on IDMs
offer great flexibility in the design and use of our proposal.

2.5   Provenance Management in Multiple Applications Context

In the previous part, we introduced the basic components of the framework in
a single provenance source context. In this part, we generalize our approach to
multiple provenance sources and address the issue of semantic correlation. Two
cases are considered: a single application having multiple provenance sources and
a set of applications having single (or multiple) provenance sources. In this context, digital objects (documents for example) are identified by different identification and authentication schemes, extra-information about the correspondence
between identifiers is needed.

In this context, an application has n different provenance sources (the log
of the http server hosting it and an application defined log for example) or n

applications have one (or several) provenance sources (for example an application producing a document, an application splitting it and an application archiving it). These provenance sources are correlated by a fusion process. Correlation is done at the DM level (compatibility) and at the instance level.

Each source or IDM is bound to a specific DM. These DMs need to be compatible together to allow to merge the different IDMs into a single one. This compatibility can be solved using algorithms for semi-automatic semantic alignment [8] but in our case, all DMs share the same root, i.e the MDM. This simplifies the alignment of DMs. More formally, we consider n $IDMs$ (IDM_1, IDM_2,..., IDM_n) with (DM_1, DM_2,..., DM_n) as associated domain models. These DMs can be aligned if they are all compatible.

Definition 6: Compatibility of DMs: Two DMs, DM_i and DM_j are compatible if there exists a generalization of DM_i called DM_{gen-i} and a generalization of DM_j called DM_{gen-j} such that: $DM_{gen-i} = DM_{gen-j}$.
Intuitively, compatibility ensures that there are no conflict between two DMs, for example a class which is specialized in a different way in the two DMs. If the DMs are not compatible, we have to simplify one (or both) IDMs using the *generalization* relationship on IDMs to cast the associated DMs to more general ones (at most the *MDM*).

Special cases of this compatibility are: (i) all the input *DMs* are the same (DM_i, DM_i,..., DM_i) and (ii) *DMs* are of two categories, MDM and a specific DM (MDM, MDM,..., DM_i, DM_i).

When all DMs are compatible, the DM_{fusion} is defined as the simple union of all DMs. Then, the IDM_{fusion} can be computed by doing the union of all triples coming from the different IDMs. Figure 4 illustrates the case of two IDMs, one defined on the MDM and the other one on a more specialized DM. The latter specializes the concept *Document* of the MDM to create the concepts *DocumentVP* (for documents with probative value) and *DocumentSVP* (for documents without probative value). The DMs are compatible and the result of the fusion is presented in Figure 5. The next step consists in correlating provenance sources at the instance level, introducing correspondence between instances from different

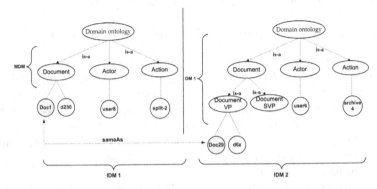

Fig. 4. An example of two compatible IDMs that could be merged

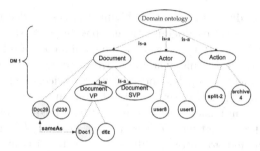

Fig. 5. The result of the fusion of two compatible IDMs

IDMs (see next subsection). Of course, this process can generate inconsistencies at the instance level that need to be addressed and solved by an administrator before generating new knowledge.

2.6   IDM Correspondence Management

As documents are processed and exchanged by different applications, it is unrealistic to imagine that a document has a unique shared identifier. Different identifiers may refer to the same document. This introduces an identification issue and requires to manage documents identification across different provenance sources. To deal with this issue, we propose to use a correspondence function. It's a user defined function, provided by the provenance provider (the person or the system providing provenance data and the import function). It permits to match identifiers corresponding to known identifiers in other provenance sources. It can be implemented by a table or by a function transforming one identifier in a source to its equivalent in another source.

The information about correspondence needs to be included into IDMs. For that, we propose to use an OWL built in property *owl:sameAs* used for ontologies mapping. It permits to specify that two URI references refer to the same thing and that these individuals have the same identity. At import time, the provenance provider adds this property to the corresponding documents to map equivalent identifiers known within other applications.

These correspondences will allow the system to infer new informations. For example, in Figure 4, *Doc1* is defined as a simple *Document* in the MDM and *Doc29* as a *DocumentVP* in *DM1*. A correspondence is defined between *Doc1* and *Doc29* and so the system can infer that *Doc1* is also a *DocumentVP*.

3   Application of the Proposed Approach to a Legal Document Archiving Use Case

3.1   Use Case Scenario

To validate our approach, we propose in this section to apply our semantic framework to a Novapost[5] use case. Novapost is a French company specialized on

---

[5] www.novapost.fr

providing collection, distribution and archiving services for electronic documents (especially human resources documents like payslips). Novapost must ensure document life cycle traceability as well as the compliance of all processing with the relative regulations. In this context, provenance sources are heterogeneous and multiple (documents reception and validation service, document distribution workflow, documents archiving service...). These sources are sparse, not easily exploitable and contain business details mandating a business expertise in order to understand and analyse them. The extraction of the full document life cycle requires to extract and go through different (may be distributed) log files. In many cases, Novapost should provide the full history of a an archived document for a customer or an auditor. Raw logs are not expressive enough or well adapted for that type of query.

The scenario of the use case is the monthly distribution of payslips of a customer company. This company uses Novapost distribution and archiving service. It sends at the end of the month a PDF file containing its employee's payslips using SFTP protocol. This document is also associated with metadata file describing the original document integrity properties. When receiving this document, Novapost performs a timestamp operation. The document is then copied. After that, the original document is archived and the copied document will be processed: the processing consists on applying a specific workflow. The specificity of the workflow depends on the input document type and the archiving policy that will be used. The workflow splits the document to generate separate payslips for each employee (in our case, this operation generates 10.000 new documents). Then, a watermark is applied to each generated document, each payslip is signed by Novapost and archived in a secure data center. Once archived, any employee can connect on the shared web portal to consult or download its payslips.

In our use case, we identify an application (Novapost application) composed by the splitting agent, the archiving agent and the distribution service. The provenance data sources for that application are the tomcat server log (catalina.out) of the server hosting it and the HTTP web server log (on which employees have to connect to manage and download their documents). We identify also the customer application sending the document. For this latter, we get from the customer company a simple log file indicating the document identifier and tracking the action of sending.

Within this use case, provenance data sources have an important size. Add to that, their raw format is not easily exploitable and information extraction and crossing is a very hard task. To deal with these issues, we will apply the proposed approach.

3.2   A Probative Value Domain Model for Novapost

Within Novapost use case, we must have probative value details about the actions performed on documents. Thus, we have to know the authentication parameters and keying material of the actors, the type and the specifities of each action, etc ... To do that, we define a domain model for Novapost by specializing the MDM as follows (see upper part of Figure 2):

1. Probative value document: described by different types of metadata with fine accuracy. This metadata can be *technical metadata* defining how the content will be interpreted and defining information about the technical process of capture, handling or restitution of the content, *structure metadata* giving the logical organization of the content like headings, paragraphs, illustrations, or *external metadata* giving descriptive information about the content like document' name, information about its integrity or its creation date...
2. Probative value actor: to ensure verifiability and accountability, we need more than the actor identifier. The probative value actor model takes into account identification and authentication parameters. These parameters can be the technical identification of the used system, the identification process (name, version) and the keying material of the actor (authentication type, key type ...).
3. Probative value actions: depending on the type of the information it changes (the document identifier or the document metadata), we have identified different probative value actions. We define an *identifying action preserving the content (IAPC)* like the action of sending the document from one system to another. We define also an *identifying action editing the content (IAEC)* like the splitting action that will create a set of documents from an original document. We define also *neutral action (NA)*: it does not alter the document metadata or identifier. The fourth action type is *structure metadata modification action (SMMA)* like the application of a watermark that does not change the document content but just its metadata. The fifth action type is *external metadata modification action (EMMA)* that modifies the external metadata of the document. The last action type is *technical metadata modification (TMMA)* that modifies the technical metadata of the document.

The Novapost domain model is described as a specialization of the MDM in the upper part of Figure 2. The bottom part represents an IDM with some instances generated by means of the import function. Figure 3 illustrates the generalization of this import IDM to the MDM level.

3.3   Provenance Queries Categories

Basically, we distinguish four categories of provenance queries. These categories are forward, backward, path and aggregation queries. They can be described as follows:

- forward query is a query identifying what output was derived from the input.
- backward query is a query addressing what input was used to generate a specific output.
- path query is a query allowing to query a path over provenance records, for example show all the documents that have follow a specific life cycle or get all the documents having a lifecycle similar to a given entry (represented as a sequence of operations).
- aggregation query is about grouping multiple values corresponding to a certain provenance criteria like counting the number of results of an operation.

We aim to test forward, backward, aggregation and path queries for the dataset presented above. Table 1 presents our benchmark queries and their characteristics.

4   Logical Architecture and Experimentations

4.1   Logical Architecture of the Provenance Management System

To support capturing, storing, managing and querying provenance, we present in this section the logical architecture of a PMS (Provenance Management System). The architecture is summarized in Figure 6, which we discuss in the rest of this section. It is based on our approach requirements to ensure genericity, extensibility as well as specific provenance management requirements.

The PMS is based on three principals components offering administration, management and querying functionalities for provenance administrators and users. The PMS can be deployed in a unique location (a kind of provenance datawarehouse) or in multiple ones (for example one per service or per cloud provider). In the second case, the multiple PMSs have to be federated to allow end-users a unified access to provenance data. We are currently working on the design of the federated architecture, but this is out of scope of this article. The modules of the PMS are:

- *Provenance management module*: offering the functionalities of importing provenance data, managing provenance sources as well as DMs, IDMs, and fusion of IDMs.
- *Provenance querying module.*
- *Provenance storage module.*

The first module, that of provenance management provides a first functionality allowing to import provenance data from raw format into the provenance store via the import function. This module offers also to the administrator the capability to manage provenance sources like adding new sources, dropping an existent one or to define a new import function. This module provides also functionalities for domain models management: add a new domain model by specializing the MDM or another DM, delete an existing DM, etc ... The provenance management module offers also IDMs generalization allowing to generate different views of the same provenance data. With this generalization, we generate automatically new IDMs corresponding to different domain models. The last functionality of this module is fusion management. This fusion is performed when provenance comes from several sources. It aims aggregating multiple sources of information to facilitate further querying and to combine them in order to generate new information that was not visible when they were separated. Thus, it is responsible on checking the semantic compatibility of the input IDMs that will be merged and to check also any inconsistencies between IDMs instances and resolve them when exists.

The second module is that of query processing which is responsible on answering end-users queries. This module takes as input a provenance query and

Table 1. Provenance query categories and their characteristics

	Provenance Query	Query description and characteristics
Query 1	What are the documents processed by Novapost's archiving module ?	Backward query restricted to a unique module of Novapost application (the archiving module).
Query 2	What happens to the document "doc-02-10-C" after the split operation was performed on it ?	Forward query after a specific operation.
Query 3	What is the number of documents generated from the document "doc02-10-C" and that are archived in users electronic vault within "CDC" third party ?	An aggregation query targeting different provenance sources.
Query 4	Was the distribution process of "doc02-10-C" successful ? If not, what are the errors detected and their causes?	We query here Novapost and customer applications. Novapost domain model is used because its accuracy level. The correlation between IDMs will tell us if the customer's distribution is in error.
Query 5	What is the full provenance of the document "doc-02-2010-s2" based on the Tomcat server log ?	A path query targeting Novapost application with a unique provenance source.
Query 6	What is the full provenance of the document "doc-02-10-s2"?	A path query within distributed logs coming from different sources. To deal with the heterogeneity of the provenance coming from those applications, we can perform a fusion process or use the MDM as a domain model. For example, we can use *owl:sameAs* to match the different document identifiers if we are using a semantic correspondence management technique.
Query 7	What is the full provenance of the document "doc-02-2010-s2" from an auditor perspective ?	This query is similar to query 6, what changes is only the domain model. Here, the system answer is different from that of Query 6 and only the actions and the documents respecting generalization rules and the target domain model ($DM_{auditor}$) will be returned.

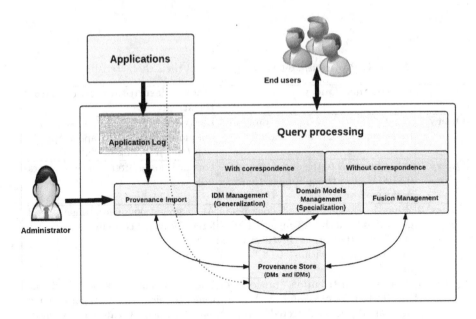

Fig. 6. Logical architecture of the Provenance Management System

returns a complete provenance chain. Query resolution can be performed according to two modes: 1) with correspondence management and 2) without correspondence management. Resolving a query without correspondence management is the traditional manner for resolving a query. However, in the second mode, the correspondences between items are considered and allow to generate new knowledge which may add new results comparing to the first mode.

The storage module is used for long term provenance storage. Physically, this module can be implemented by any storage technology. It can be a relational DBMS, an RDF triple store, a NoSQL DBMS, etc... But of course, the choice of a specific storage system will constraint the supported query language. The provenance store module can offer two APIs. One corresponding to the use of import functions and the other one to the direct storage of application logs if they are natively formatted with our models.

4.2   Implementation, Experimentations and Scalability

In this part, we validate our approach by using real Novapost provenance data. The experiments we performed aim validating the feasibility of the proposed approach and to test its performance and scalability.

We evaluate two implementations of our framework. The first one uses Sesame RDF store as a storage system. It supports SPARQL as its native query language but experiments show that it doesn't scale. Problems of efficiency and scalability of Sesame have already been observed and several proposals (e.g [9,10]) have been proposed to alleviate these issues but have not an adequate level of maturity and that is the reason why we have chosen Sesame for our experiments. Relational

DBMSs are an alternative solution but we prefer to use a NoSQL system for three main reasons: (i) SQL is limited compared to map-reduce style query language, (ii) NoSQL systems allow to easily program parallel queries on multiple servers allowing vertical scalability and (iii) we need to store schema variable data and so NoSQL systems are naturally suitable due to their no schema approach. So, in the second experimentation, we use a NoSQL DBMS, coupled with a map-reduce based query interface. Of course, we cannot use SPARQL queries in this case, but all our benchmark queries are expressible and experiments show that it scales. Experiments were performed on a real dataset composed of two provenance sources coming from Novapost: the log of Tomcat server having 850 MB size and the log of the Apache http server having about 250 MB size. These two provenance sources contain traces of processing operations, distribution and archiving of documents of four months for about twenty customers and a total of 1 million pdf documents, mainly payslips. This is about 2.5 million log lines that must be analyzed and integrated into the storage module.

All experiments were conducted on a DELL workstation having the following configuration: Intel Xeon W3530 2.8GHz CPU; 12GB DDR2 667 RAM; 220 GB (10,000 rpm) SATA2 hard disks and Fedora 13 (Goddard) 64-bit as operating system. We installed the following software on that workstation:

1. Sesame[6] Version 2.6.3 (Native Storage). This version of Sesame is characterized also by its support to the SPARQL 1.1 Graph Store HTTP Protocol. We have deployed our Sesame as a Java Servlet application in Apache Tomcat.
2. Apache Tomcat (Version 6.0.35).
3. Java (JRE v1.6.0_18).
4. CouchDB 1.1.0.
5. BigCouch 0.4a.

4.3   Experiments with Sesame, an RDF Triple Store

A triple store is a system providing a mechanism for storage and access to RDF graphs. We choose Sesame which is an open source framework for storing, inferencing and querying RDF data. Sesame provides a connection API, inferencing support, availability of a web server and SPARQL endpoint. It provides also a support for multiple backends like MySQL and PostgreSQL.

Representing Provenance in Sesame. We have used python RDFLib[7] to develop the import function. It parses the input logs and creates the RDF graphs. These graphs are stored into RDF files. Once created, these RDF files are uploaded into the Sesame Native store (native triple stores provide persistent storage with their own implementation of the database) via Sesame's command line tool.

We have used Novapost domain model as an import model (DM_{import}). Thus, a document instance generates an average of seven triples, an action seven triples too and an actor six triples. Listing 1.1 presents an RDF sample of provenance data created by the import function and stored into Sesame.

---

[6] www.openrdf.org
[7] www.rdflib.org

Listing 1.1. A sample of the RDF provenance data stored in Sesame

```
<nova:AuthenticatedActor  rdf:about=http://www.track.novapost.fr/
    instantiated/actors/sftpSender>
<nova:hasId>sftpSender</nova:hasId>
<nova:type>AuthenticatedActor</nova:type>
<nova:hasActorId>sftpSender</nova:hasActorId>
</nova:AuthenticatedActor>
<nova:ProbativeValueAction  rdf:about=http://www.track.novapost.fr/
    instantiated/actions/send28-11-2011>
<nova:hasInput  rdf:resource=http://www.track.novapost.fr/
    instantiated/documents/1423438/>
<nova:precededBy>emit-63</nova:precededBy>
<nova:performedAt rdf:datatype=http://www.w3.org/2001/XMLSchema#
    dateTime>2011-11-28T21:32:52</nova:performedAt>
<nova:performedBy>sftpSender</nova:performedBy>
<nova:hasActionId>send-28-11-2011</nova:hasActionId>
<nova:hasOutput  rdf:resource=http://www.track.novapost.fr/
    instantiated/documents/744439/>
</nova:ProbativeValueAction>
<nova:ProbativeValueDocument  rdf:about=http://www.track.novapost.fr
    /instantiated/documents/1423438>
<nova:hasApplicationId>sftpSender</nova:hasApplicationId>
<nova:hasStatus>sent</nova:hasStatus>
<nova:hasDocumentId>doc1423438</nova:hasDocumentId>
<nova:hasSystemId>hrAccess</nova:hasSystemId>
<nova:hasSHA256FingerPrint>
    D12AB8F6200BD0AAE1B1F5B9B5317F8F4113B2B9C015B3734045FA463B5A6D0D
    </nova:hasSHA256FingerPrint>
</nova:ProbativeValueDocument>
```

SPARQL Queries over Provenance Data. The different provenance queries were formulated quite easily in SPARQL. Each query targets an IDM what implies that the domain model is already fixed. This domain model is specified using the keyword PREFIX in the SPARQL query header. For path queries (Q5,Q6 and Q7), SPARQL is able to answer recursive queries and to create the list of the documents involved in the lifecycle of a given document. We present above as examples, three SPARQL queries of our benchmark:

Listing 1.2. Q1 in SPARQL

```
PREFIX nova: <http://track.fr/DM/novapost/probative.owl>
SELECT   ?documents
FROM     <http://track.fr/applications/novapost.rdf>
WHERE { ?actions nova:performedBy ''nova-archiverAgent''.
                 ?actions nova:hasInput ?documents.
}
```

Listing 1.3. Q3 in SPARQL

```
PREFIX nova: <http://track.fr/DM/novapost/probative.owl>
SELECT (count(?documents) as ?c)
FROM     <http://track.fr/applications/novapost.rdf>
WHERE { ?actions   nova:hasInput ''doc02-10-C''
        ?actions   nova:hasOutput ?documents.
        ?documents  nova:hasStatus ''archived''.
               }
```

Table 2. Arithmetic mean of queries execution time for the different datasets in Sesame (in milliseconds)

	50 K	250 K	1 M	5 M	24 M
Q1	26	36	49	63	439
Q2	23	92	464	2463	13174
Q3	51	52	55	61	173
Q4	15	17	19	21	994
Q5	157	602	3020	16014	85647
Q6	162	3116	1322	16524	88290
Q7	155	593	2980	15810	84457

Listing 1.4. Q6 in SPARQL

```
PREFIX nova: <http://track.novapost.fr/DM/mdm.owl>
SELECT ?linkedAction FROM <http://track.fr/applications/novapost.
    rdf>
WHERE { ?action nova:hasOutput ''doc-02-10-C''.
            ?action (nova:hasInput/^nova:hasOutput)* ?linkedAction.
        }
```

Experiments with Sesame. Our import function generated approximately 24 million triples (exactly 24.000.416 triples) that we have queried using SPARQL. The loading of provenance data into Sesame shows that load performance is only good for small datasets (less than 1M triples).

Table 2 illustrates the results of the execution time of our queries. For every query, the shown result is the arithmetic mean of the query execution time over 10 runs per query.

When we conducted the tests on our largest dataset (24M), we encountered memory time-out problems. By increasing the java Xmx size until 6 GB, we have succeed to reduce enormously this problem but we do not eliminate it completely (4 time-outs on the set of 70 tests for the 24M dataset). These results allow us to draw several interpretations:

- for small datasets, response times are satisfactory even if the query involves pattern matching or path computation.
- aggregation and path queries are the slowest queries (compared to simple selection or filter queries).
- all queries depend on data volume and response size. For the largest dataset, this impact is very remarkable, especially on aggregation and path queries.
- Sesame does not provide linear query response time and will not scale for large datasets and complex queries (like Q5, Q6 and Q7) .

4.4   Experiments with CouchDB, a NoSQL DBMS

NoSQL DBMS is a new word referring to different types of non relational DBMS, that is key-value DBMS, document DBMS and graph DBMS. They have been gained lots of popularity these days with their use by big web players such as Google (BigTable) or Facebook (Cassandra) for example. Many of these NoSQL DBMS are coupled with a data parallel model called map/reduce [11] initially designed by Google and popularize by its open-source implementation Hadoop. Map-reduce allows to evaluate complex processing on large datasets using parallelism with a simple paradigm (map which maps data to nodes and reduce which aggregates the result produced by the nodes). Even if there are many discussions on the effectiveness of NoSQL DBMS related to parallel DBMS [12], NoSQL DBMS seem to provide an efficient and cheap solution to handle queries on large datasets [13]. In our context, we believe that the use of NoSQL database for large scale provenance management is interesting for the following reasons:

1. Provenance data structure is flexible and varies from an application to another. This corresponds very well to the schema free characteristic of NoSQL databases.
2. Provenance data does not need transactional guarantees when integrated.
3. Provenance data is static, never updated and is used in an append only mode (such is the case for data analytics).
4. Provenance data is characterized by its large volumes what implies efficient query techniques adapted for big data.
5. Some NoSQL databases provide querying techniques based on map/reduce what allows efficient selection and aggregation of data thanks to parallelism.

We have chosen CouchDB which is a document oriented NoSQL database written in Erlang. The approach of CouchDB is to use the suitable web technologies such as Representational State Transfer (REST) [14], JavaScript Object Notation (JSON) as a data interchange format and the ability to integrate with infrastructure components (such as load balancers and caching proxies, etc...) to build a document database from scratch. CouchDB is distributed and able to replicate between server nodes as well as clients and servers incrementally [15]. JavaScript (or Erlang) functions select and aggregate documents and representations of them in a map/reduce manner to build views of the database which also get indexed.

Representing and Querying Provenance in CouchDB. We have used the same real datasets and we have developed a new import function in Java. This function uses Ektrop[8] which is a CouchDB Java client facilitating the interaction with CouchDB and providing insert, bulk insert and querying functionalities.

We have chosen to represent instances of IDM as three distinct document types in CouchDB: **ActorDocument**, **ActionDocument** and **Document**. Indeed, this typing is used just to project the basic MDM concepts. It imposes no relationship or constraint among data. Each type may refer other documents of any other type. This typing has been defined in order to consider the nature of the entities of MDM as first-order elements. This allows to express and execute

---

[8] www.ektorp.org

queries in a more natural and more efficient manner. The structure of provenance documents is illustrated in Listing 1.5.

As the triple logic of Sesame is different from CouchDB document-oriented logic, the total number of the documents generated by our import function is different from the tuples number mentioned above (24M tuples). Our import function generated approximately 3M (exactly 3.000.517) documents. To insert them, we have used the bulk insert feature provided by Ektorp. This function is especially interesting for large datasets.

Listing 1.5. JSON representation of provenance documents stored in CouchDB

```
Action Document:
{
    "_id": "001e38d82a5818ea60dee3cef1001525",
    "_rev": "1-0d7086883d47c872d5c5376113d11ec4",
    "name": "SEND",
    "id": "send-28-11-2011",
    "type": "ActionDocument",
    "timestamp": "28/11/2011-21:32:52",
    "input": [
        "744438"
    ],
    "output": [
        "744439"
    ],
    "actor": "sftpSender",
    "precededBy": "emit-63"
}

Actor Document:
{
    "_id": "001e38d82a5818ea60dee3cef103794c",
    "_rev": "1-702717b4f4266998791c637abdd6ea4b",
    "name": "sftpSender",
    "type": "ActorDocument",
    "characteristic": "Authenticated"
}

Document:
{
    "_id": "001e38d82a5818ea60dee3cef1032d1c",
    "_rev": "1-2dee8a50897071be3140dabac4290233",
    "type": "Document",
    "status": "sent",
    "ids": [ "cegid:744438","nova:paie\_11-2011-custCE4"
    ],
    "path":[''output,split,744439,28/11/2011-21:32:55''],
    "sha256": "D12AB8F6200BD0AAE1B1F5B9B5317F8F4113B2B9C015B3734045FA463B5A6D0D
        ",
    "identifier": "744438"
}
```

Once our provenance data created and loaded into CouchDB, we should define the views permitting to query it. Views are the primary tool used for querying CouchDB databases. They can be defined in JavaScript (although there are other query servers available). Two different kinds of views exist in CouchDB: permanent and temporary views. Temporary views are not stored in the database, but rather executed on demand. This kind of query is very expensive to compute each time they get called and they get increasingly slower the more data you have in a database. Permanent views are stored inside special documents called design documents. This document contains different views over the database. Each view is identified by a unique name inside the design document and should define a map function and optionally a reduce function:

– The map function looks at all documents in CouchDB separately one after the other and creates a map result. The map result is an ordered list of key/value pairs. Both key and value can be specified by the user writing the map function. A map function may call the built-in emit(key, value) function 0 to N times per document, creating a row in the map result per invocation (a sample of the result of the execution of a map function for Q1 is listed in 1.9).

- The reduce function is similar to aggregate functions in SQL, it computes a value over multiple documents. This function operates on the output of the map function and returns a value.

The following listings (listing 1.6, 1.7 and 1.8) presents three views among the views we have defined to query CouchDB. These views are the translations of the SPARQL queries listed in part 4.3. Even if these expressions are not too complex to define and to understand, it is not as declarative and readable as SPARQL queries.

Listing 1.6. Q1: Backward query

```
''documents_by_actor'': {
        ''map'': ''function(doc)
            {if((doc.actor) && (doc.type=='ActionDocument'))
                emit(doc.actor, doc._id)}''
    }
```

Listing 1.7. Q3: Aggregation query

```
''documents_archived_number'': {
        ''map'': ''function(doc) {if((doc.type=='Document') && (doc
            .status) && (doc.archiver) &&(doc.identifier))
                emit([doc.identifier,doc.status,doc.
                    archiver], doc._id)}'',
        ''reduce'': ''_count''
    }
```

Listing 1.8. Q6: Path query

```
''path_by_docid'': {

if (doc.identifier) {
    emit([doc.identifier, 0], doc.path);
    if (doc.inputs) {
        for (var i in doc.inputs) {
            emit([doc.identifier, Number(i)+1], {path: doc.path[i]});
        }
    }
}

if (doc.outputs) {
    for (var j in doc.outputs) {
        emit([doc.identifier, Number(j)+1], {path: doc.path[j]});
    }
    }
  }
}
```

Listing 1.9. A sample of the result of the execution of a map function for Q1

```
{"id":"8447c2517d6bbf56b02523c3a9d74c6e","key":{"identifier":"
    UBISOFT:","localId":"","identifcationSystemId":"UBISOFT"},"
    value":"8447c2517d6bbf56b02523c3a9d74c6e"},
{"id":"84c9764b26b5951e57f1fda6542ca8aa","key":{"identifier":"
    UBISOFT:","localId":"","identifcationSystemId":"UBISOFT"},"
    value":"84c9764b26b5951e57f1fda6542ca8aa"},
{"id":"8659596bc94d33528c37b472dcfeab10","key":{"identifier":"
    UBISOFT:","localId":"","identifcationSystemId":"UBISOFT"},"
    value":"8659596bc94d33528c37b472dcfeab10"},
{"id":"95ddc9f06131f02e762230658b68cab4","key":{"identifier":"
    UBISOFT:","localId":"","identifcationSystemId":"UBISOFT"},"
    value":"95ddc9f06131f02e762230658b68cab4"},
{"id":"98ac0810d6a85438dc776d06b87b9b18","key":{"identifier":"
    UBISOFT:","localId":"","identifcationSystemId":"UBISOFT"},"
    value":"98ac0810d6a85438dc776d06b87b9b18"},
{"id":"a01d8a53ca64d37279dfb0f5a6002445","key":{"identifier":"
    UBISOFT:","localId":"","identifcationSystemId":"UBISOFT"},"
    value":"a01d8a53ca64d37279dfb0f5a6002445"},
{"id":"b36addfb79f83fa4b7eea53c5705c65a","key":{"identifier":"
    UBISOFT:","localId":"","identifcationSystemId":"UBISOFT"},"
    value":"b36addfb79f83fa4b7eea53c5705c65a"},
{"id":"cdb8d8ad9fdcd42be4296c4532d87e46","key":{"identifier":"
    UBISOFT:","localId":"","identifcationSystemId":"UBISOFT"},"
    value":"cdb8d8ad9fdcd42be4296c4532d87e46"},
{"id":"dccdba34a36fcb5cb26231c3747cfb49","key":{"identifier":"
    UBISOFT:","localId":"","identifcationSystemId":"UBISOFT"},"
    value":"dccdba34a36fcb5cb26231c3747cfb49"},
```

Experiments with CouchDB. Table 3 illustrates the average queries run time on CouchDB. These results allowed us to draw the following remarks:

- the first query execution takes a long time because it triggers database indexing. We have not taken into account the time required for this query execution in our measurements.
- forward and backward queries are quite efficient while aggregation queries are an order of magnitude slower.
- path queries are not so efficient but the execution is still convenient.
- the execution time of all queries is independent from the dataset size. Map/reduce queries are efficient on small and large dataset.
- view definition (the definition of map/reduce function) for provenance queries is not simple and declarative as SPARQL. For example, path queries need recursion. Such feature is not currently integrated into CouchDB. For this type of queries, additional attributes must be integrated into the data model to link provenance data. In our case, we added to Action Documents a field called "path". This field keeps the idenfiers of all the actions using the current action as input or output.

Provenance Data Scalability. We aim to build a scalable provenance management system. We have succeed to insert 15M documents into a single CouchDB node (by running the same provenance data integration test 5 times) but we encountered memory problems while performing the database indexation. Instead of tuning a single node, we have created a CouchDB Cluster and have distributed

Table 3. Arithmetic mean of queries execution time for the different datasets in CouchDB (in milliseconds)

	50 K	250 K	1 M	3 M
Q1	57	58	61	63
Q2	48	48	49	51
Q3	561	565	566	566
Q4	55	55	57	58
Q5	4106	4227	4325	4331
Q6	4107	4231	4328	4333
Q7	4120	4280	4282	4285

the database storage and indexation. For that, we have used BigCouch[9] which is an open source system provided by Cloudant permitting to use CouchDB as a cluster.

To analyse the response time of large data volumes, we have configured a three machines cluster (having the same hardware configuration of that mentioned in 4.2. We create a distributed database on which we have integrated respectively 15 (dataset D1) and 30 (dataset D2) million provenance documents (i.e CouchDB documents). These experiments aim to test the impact of data size on provenance queries. To do that, we run our provenance queries and we measure the average runtime over 10 executions for datasets D1 and D2. The results are illustrated in Table 4 and show that the execution time is still independent of the size of the datasets due to the distribution of the query load on the different servers.

Table 4. Average query runtime on BigCouch/CouchDB for the datasets D1 et D2 (in milliseconds)

Query	Q1	Q2	Q3	Q4	Q5	Q6	Q7
Execution time for D1 (15M)	60	51	558	59	4339	4345	4343
Execution time for D2 (30M)	63	53	560	60	4340	4347	4346

The last part of our experiments is focused on path queries. These queries compute intermediate results in a recursive manner. We want to analyze the impact of the size of the intermediate results on these queries. We want to know if the response time of these queries is related only to the number of intermediate results and not the amount of data. For this, we perform tests on Q7 with 3 (dataset D'1), 15 (dataset D'2) and 30 (dataset D'3) million provenance documents respectively. For each dataset, the first test involves six intermediate results, the second 15 and the third 20. The results in Table 5 show that the size of the intermediate results has a significant impact on the execution time. An increase of 333% in size implies an increase of 215% in execution time.

---

[9] http://bigcouch.cloudant.com

Table 5. Impact of the size of intermediate results on the query Q7 (in milliseconds)

Number of intermediate results	6	15	20
Execution time for D'1	4285	6829	9374
Execution time for D'2	4343	6835	9387
Execution time for D'3	4359	6829	9397

4.5 Comparison of Sesame and CouchDB Experiments

Provenance Data Loading. Figure 7 illustrates the time required to load the datasets as RDF files created by the import script as well as the time required to integrate provenance documents into CouchDB. For small datasets (less than 1 million triple), Sesame is fast and the load time was reasonable. For larger data, Sesame is slow and the loading of all the triples took more than 7 hours (28302 seconds). For CouchDB, the bulk insert function allows to integrate large data volumes in a linear load time.

Provenance Querying. By analysing the tables of queries execution time for Sesame (Cf. Table 2) and CouchDB (Cf. Table 3), we notice that for the different queries, Sesame shows a good performance for small datasets that exceeds that of CouchDB especially for complex queries. However, when increasing provenance data volume, this trend changes and CouchDB performs much better.

Also, Table 3 illustrates that CouchDB presents almost constant time. Also, it is clear that for large datasets, CouchDB presents a very good performance compared to Sesame for aggregation queries (Q3) and path queries (Q5, Q6 and

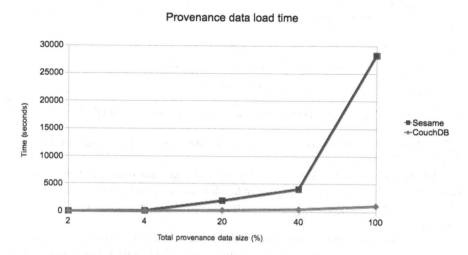

Fig. 7. Provenance data load time into Sesame and CouchDB (in seconds)

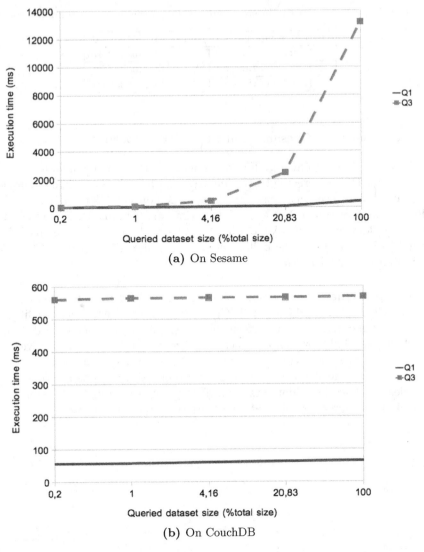

(a) On Sesame

(b) On CouchDB

Fig. 8. Evolution of query response time for Q1 and Q3

Q7). These interpretations are better illustrated in the following graphs (Cf. Fig.8 and Fig.9).

Also, the tests performed on BigCouch and that are illustrated in Table 4 show that CouchDB scales in a linear fashion for all provenance queries. So, if we want to store larger provenance data, we have just to add a new server to the cluster.

Nevertheless, one can notice that the per node scalability is very poor since we have not succeed in storing and querying around 50 Gigabytes per node. This low efficiency seems related to a problem of configuration of CouchDB and we

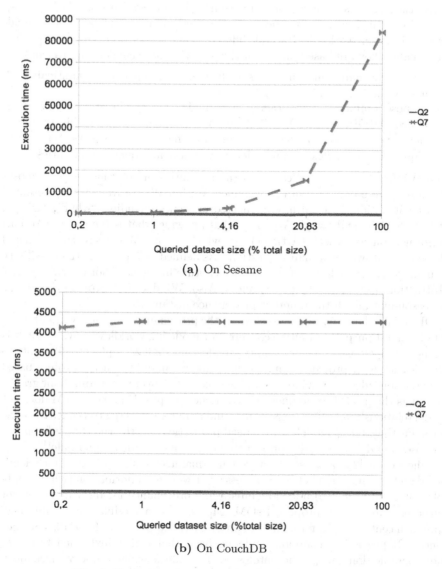

(a) On Sesame

(b) On CouchDB

Fig. 9. Evolution of query response time for Q2 and Q7

have to investigate in this direction to be able to really demonstrate that our system scales.

5   Related Works

Provenance challenges and opportunities has been the target of different research works [16,17]. Many approaches center on a "workflow engine perspective of any system" and consider that operations are orchestrated by that workflow engine.

This vision ignores what is domain specific relevant (semantics, relationships). To address this issue, provenance modeling and management aspects were addressed in other works [18,19] and different frameworks were proposed [20,21]. Principally, they are based on one of the following management approaches:

- group provenance and business specificities in a unique and semantically rich model like W7 [22].
- propose a minimal, semantically poor model that can be annotated and enriched like OPM [7], PrOM [23,24] or Hartig's model [25]. For this approach, the enrichment techniques of the minimal model are the keys. Such techniques should be flexible and offer easy collect and query possibilities.

The W7 [22] model is dedicated to the semantics of provenance, it proposes an ontological model for data provenance. Within W7, provenance is presented as a combination of seven interconnected elements including "what", "when", "who", "which" and "why". We argue that knowing what is the action (Action) performed on an object (Artifact), by which actor (Actor) are the principal elements to answer provenance questions. As defined, W7 provides the possibility to manage the granularity of provenance capture but it does not provide queries abstraction over the captured instances. Also, W7 does not provide the ability to reason and match the different provenance instances.

OPM [7] is the result of a joint effort of the provenance community. It defined a set of general-purpose primitive concepts for modeling workflow provenance. It is a generic and interoperable provenance model that shall be applied to different application areas. It provides semantic enrichment capabilities using profiles based on extension rules. It provides reasoning capabilities over OPM compliant graphs as well as the possibility to query that enriched graphs. However, OPM does not model relationships as first class entities and requires use of tags on edges to define roles. Further, OPM models only causal properties and the rules for inferencing proposed in OPM are easily contradicted due to its generic graph model[10].

Sahoo et al [23] proposes a provenance management approach for eScience based on the separation between expressive provenance information (what is only relevant to provenance) and domain-specific provenance for data management purposes. Later, they propose PrOM [24] which is a modular, multi-ontology approach centered on a foundational ontology called *provenir*. PrOM is eScience centric. It provides a common modeling basis that can be extended to create interoperable domain-specific ontologies. To build a scalable query mechanism supporting complex queries, PrOM defines a classification of queries according to their characteristics which are: retrieving provenance information, retrieving data entities that satisfy provenance and retrieving operations on provenance information. These operators are designed as a Java-based API to support queries over RDF data store.

In the context of open data, Hartig defined a provenance model [25] that is a specialization of OPM [26]. His model focuses on two dimensions: data creation

---

[10] http://twiki.ipaw.info/pub/Challenge/ OpenProvenanceModelWorkshop/ FeedbackonOPM.pptx, retrieved on Dec 21, 2011.

and data access of RDF data. The genericity of the model gives applications the choice to refine it according to their use cases. However, we argue that the proposed approach is strongly bound to the context of linked data.

The provenance community is increasingly interested in storage and querying aspects of provenance data. Chebotko et al [27] proposed RDFPROV, a relational RDF store for querying and managing scientific workflow provenance. This semantic web driven system is optimized for querying and managing scientific workflow provenance metadata. The architecture of RDFPROV seamlessly integrates the interoperability, extensibility, and reasoning advantages of semantic web technologies with the storage and querying power of an RDBMS. To support this integration, three model mappings were introduced that are schema mapping, data mapping and SPARQL-to-SQL query mapping. The experimental study of RDFPROV showed that the proposed mapping algorithms are efficient and scalable. Compared with existing general-purpose RDF stores like Jena, Sesame, AllegroGraph and BigOWLIM, RDFPROV provides an improved efficiency provenance metadata management. However, RDFPROV don't address the questions about provenance data loading as well as scalability issues with too many millions of provenance instances.

In the same category, Groth et al [20] proposed within the European project Provenance[11] a logical architecture for provenance systems. This architecture describes a provenance model structure based on p-assertions. Security and scalability aspects are considered as well: the first one provides the details about secure transmission and access control of provenance to provenance stores, and a series of scenarios are given to illustrate how different modes of interaction with the secured system will take place. For the scalability aspect, the need for distribution of provenance stores is emphasized, and a set of deployment patterns for recording process documentation into distributed stores is given. Zhao et al [28] have proposed an approach for querying provenance in distributed environments. This approach is based on a provenance index service keeping the mappings between the provenance repositories and the data artifacts. This approach considers just two categories of provenance queries.

In our work, we have addressed different dimensions of provenance management within a global approach aiming to build a scalable semantic provenance management system. Unlike W7 [22], we proposed to handle provenance modeling using an OPM-like [7] vision based on a minimal and semantically poor model that can be enriched (the same logic has been adapted by PrOM [24] and Hartig [25]). However, the genericity of these models leads sometime to inconsistencies between the enriched provenance models. To avoid this issue, we have proposed a set of strict inheritance rules permitting to define new provenance models from the minimal model (MDM) defined as a subset of OPM. Zhao et al [28] have proposed a system for querying provenance in distributed environments. It is based on an index service deployed on a mediator. This later analyses the provenance queries and query the suitable provenance repository based on repository-artifact metadata. We believe that this approach lacks correlation

---

[11] www.gridprovenance.org

between sources and may influence the relevance of the response. To deal with the distributed aspect of provenance sources, we have proposed a fusion process between instantiated domain models (IDMs) preceded by a semantic compatibility check (at models levels). This fusion is followed by an enrichment process based on correspondence management between distributed provenance data. Thus, we create new knowledge and enrich IDMs that can be queried even if the initial schema does not produce any response. Another contribution of our work is that of automatic IDMs generation from the IDM_{import}. This technique permits to create different business visions over the same log.

Like Growth et al [20], we have proposed a logical architecture of a provenance management system. Both of these architectures are bound to a whole provenance management approach taking into account collection, distribution, storage and querying aspects of provenance. Growth's architecture proposes how to build a PMS form scratch and defines even how applications should structure their provenance data and how to deliver it to the provenance collection component. However, our architecture is based on the use of already existent logs and permits to make any application provenance aware. Despite this advantage (making any application provenance aware), our architecture is not intrusive because it does not require the modification of the application source code.

We have also tested and analysed scalability aspects of provenance storage and querying. Chebotko et al [27] have proposed RDFPROV, a relational RDF store based on a set of mappings (schema mapping, data mapping, and SPARQL-to-SQL) to combine semantic web technologies advantages with the storage and querying power of RDBMs. However, RDFPROV do not address the questions about provenance data loading as well as scalability. Similarly, our work combines two type of technologies: semantic web for provenance modeling and correlation and NoSQL/map-reduce for provenance querying and storage. However, our work have also addressed provenance data loading and querying issues and have provided different interpretations about provenance queries expressiveness and scalability. These interpretations will permit to PMS designers to justify their technical choices according to their expressiveness and scalability needs.

6   Conclusion

As data is shared across networks and exceeds traditional boundaries, provenance becomes more and more important. By knowing data provenance, we can judge its quality and measure its trustworthiness. Today, new provenance requirements arise and jeopardize the existent provenance models for single applications. In a distributed environment, these models are not yet in a stage of maturity and lack separation between what is provenance specific and domain specific. In this paper, we propose to use the power of semantic web technologies for heterogeneous, multiple sources and decentralized provenance integration. Based on RDF, OWL and SPARQL, we have presented a semantic provenance framework permitting to capture provenance from different heterogeneous sources, to enrich it semantically, to correlate and match it with other sources to provide rich answers about the whole documents life cycle. Using RDF facilitates data integration across boundaries and solves syntactic heterogeneity issues. Coupled with the use of a generic and minimal domain model (MDM), *specialization and generalization* relationships, we address semantic interoperability issues. To model business specificities, OWL ontologies are used to define semantically rich domain models what permits to construct high level representation of provenance. The correlation between instantiated ontologies allows us to link different provenance sources and to generate new knowledge. Regarding query, the use of the SPARQL standard permits to define complex and domain specific queries over provenance data.

Although presented in the context of electronic document management, our approach is generic and can be applied to any type of log file. However, this approach is more interesting for semantically rich logs.

In the second part of this article, we have focused on provenance querying and storage. First, we have presented the storage structure of provenance data in Sesame, an RDF triple store. This storage technology shows important limitations in term of scalability. To alleviate scalability issue, we replace the PMS storage module with a NoSQL document-oriented database (CouchDB). We propose a document storage structure on CouchDB that we evaluate and compare with Sesame. CouchDB scales up linearly for the different queries. Also, it presents good performance for aggregation and path queries. However, per node efficiency is not very good and we have to tune CouchDB configuration to improve it.

As on-going works, we are working on the unified federation of PMSs, corresponding to different service or cloud providers. The idea here is to use a mediator-based architecture on top of PMSs. The mediator will be in charge of the correlation between provenance sources and of the query rewriting on the PSMs. Specific rewriting methods can be used because this mediator will have a minimal knowledge of all the instances and so can achieve better optimizations. We have also to go further in the design of the correlation between sources, studying the type of inference that can be computed.

References

1. Goble, C.: Position statement: Musings on provenance, workflow and (semantic web) annotations for bioinformatics. In: Workshop on Data Provenance and Derivation (October 2002)
2. Curbera, F., Doganata, Y., Martens, A., Mukhi, N.K., Slominski, A.: Business Provenance – A Technology to Increase Traceability of End-to-End Operations. In: Meersman, R., Tari, Z. (eds.) OTM 2008, Part I. LNCS, vol. 5331, pp. 100–119. Springer, Heidelberg (2008)
3. Agrawal, P., Benjelloun, O., Sarma, A.D., Hayworth, C., Nabar, C., Sugihara, T., Widom, J.: Trio: A system for data, uncertainty, and lineage. In: Very Large Data Bases, pp. 1151–1154 (2006)
4. Hasan, R., Sion, R., Winslett, M.: Preventing history forgery with secure provenance. ACM Transactions on Storage 5, 12:1–12:43 (2009)

5. Sakka, M.A., Defude, B., Tellez, J.: Document Provenance in the Cloud: Constraints and Challenges. In: Aagesen, F.A., Knapskog, S.J. (eds.) EUNICE 2010. LNCS, vol. 6164, pp. 107–117. Springer, Heidelberg (2010)
6. Kiran-Kumar, M.R., Margo, S.: Provenance as first class cloud data. SIGOPS Oper. Syst. Rev. 43, 11–16 (2010)
7. Moreau, L., Clifford, B., Freire, J., Futrelle, J., Gil, Y., Groth, P., Kwasnikowska, N., Miles, S., Missier, P., Myers, J., Plale, B., Simmhan, Y.L., Stephan, E., Bussche, J.V.: The open provenance model core specification (v1.1). Future Generation Computer Systems (July 2010)
8. Euzenat, J., Shvaiko, P.: Ontology matching. Springer, Heidelberg (2007)
9. Liu, X., Thomsen, C., Pedersen, T.B.: 3XL: Supporting efficient operations on very large OWL Lite triple-stores. Information Systems (December 2010)
10. Neumann, T., Weikum, G.: Rdf-3x: a risc-style engine for rdf. Proc. VLDB Endow. 1(1), 647–659 (2008)
11. Dean, J., Ghemawat, S.: Mapreduce: a flexible data processing tool. Commun. ACM 53(1), 72–77 (2010)
12. Stonebraker, M., Abadi, D., DeWitt, D.J., Madden, S., Paulson, E., Pavlo, A., Rasin, A.: Mapreduce and parallel dbmss: friends or foes? Commun. ACM 53(1), 64–71 (2010)
13. Pavlo, A., Paulson, E., Rasin, A., Abadi, D.J., DeWitt, D.J., Madden, S., Stonebraker, M.: A comparison of approaches to large-scale data analysis. In: Proceedings of the 35th SIGMOD International Conference on Management of Data, SIGMOD 2009, pp. 165–178. ACM, New York (2009)
14. Fielding, R.T.: Architectural styles and the design of network-based software architectures. PhD thesis (2000)
15. Software Foundation Apache: Apache couchdb: introduction (2008-2010), http://couchdb.apache.org/docs/intro.html
16. Kiran Kumar, M.R.: Foundations for Provenance-Aware Systems. PhD thesis, Harvard University (2010)
17. Davidson, S.B., Freire, J.: Provenance and scientific workflows: challenges and opportunities. In: Proceedings of ACM SIGMOD, pp. 1345–1350 (2008)
18. Simmhan, Y.L., Plale, B., Gannon, D.: A survey of data provenance in e-science. SIGMOD Rec. 34, 31–36 (2005)
19. Freire, J., Koop, D., Santos, E., Silva, C.T.: Provenance for computational tasks: A survey. Computing in Science and Engineering, 11–21 (2008)
20. Groth, P., Jiang, S., Miles, S., Munroe, S., Tan, V., Tsasakou, S., Moreau, L.: An architecture for provenance systems. Technical report (February 2006), http://eprints.ecs.soton.ac.uk/13196 (access on December 2011)
21. Freitas, A., Legendre, A., O'Riain, S., Curry, E.: Prov4j: A semantic web framework for generic provenance management. In: The Second International Workshop on Role of Semantic Web in Provenance Management, SWPM 2010 (2010)
22. Ram, S., Liu, J.: A new perspective on semantics of data provenance. In: The First International Workshop on Role of Semantic Web in Provenance Management, SWPM 2009 (2009)
23. Sahoo, S.S., Sheth, A., Henson, C.: Semantic provenance for escience: Managing the deluge of scientific data. IEEE Internet Computing 12, 46–54 (2008)
24. Sahoo, S.S., Barga, R., Sheth, A., Thirunarayan, K., Hitzler, P.: Prom: A semantic web framework for provenance management in science. Technical Report KNOESIS-TR-2009, Kno.e.sis Center (2009)

25. Hartig, O.: Provenance information in the web of data. In: Second Workshop on Linked Data on the Web, LDOW (2009)
26. Moreau, L.: The foundations for provenance on the web. Found. Trends Web Sci. 2, 99–241 (2010)
27. Chebotko, A., Lu, S., Fei, X., Fotouhi, F.: Rdfprov: A relational rdf store for querying and managing scientific workflow provenance. Data Knowl. Eng., 836–865 (2010)
28. Zhao, J., Simmhan, Y., Gomadam, K., Prasanna, V.K.: Querying provenance information in distributed environments. International Journal of Computers and Their Applications (IJCA) 18(3), 196–215 (2011)

A Unified Conceptual Framework
for Service-Oriented Computing
Aligning Models of Architecture and Utilization

Colin Atkinson[1], Philipp Bostan[1], and Dirk Draheim[2]

[1] University of Mannheim
atkinson@informatik.uni-mannheim.de, bostan@uni-mannheim.de
[2] University of Innsbruck
draheim@acm.org

Abstract. Given the importance of clients in service-oriented computing, and the ongoing evolution of distributed system realization technologies from client/service architectures, through distributed-object and
service-oriented architectures to cloud computing, there is a growing need
to lower the complexities and barriers involved in the development of
client applications. These range from large scale business applications
and business processes to laptop programs and small "apps" on mobile devices. In this paper we present a unified conceptual framework
in which the basic concerns and viewpoints relevant for building clients
of service-oriented, distributed systems can be expressed and related to
one another in a platform-independent, non-proprietary way. The basic
concerns used to structure the framework are the level of abstraction
at which a system is represented and the roles from which the software
entities of a distributed system are viewed. Using the various concepts
and models supported in the framework it is possible to customize and
simplify each client developer's view and to simplify the way in which
service providers develop and maintain their services. This paper provides an overview of the framework's foundations and concepts. We also
present the vision behind this conceptual framework and present a small
example to show how the models contained in the framework are applied
in practice.

Keywords: distributed computing, model-driven development, service-
oriented architectures.

1   Introduction

Since distributed computing became a viable solution for building enterprise
systems in the late 1980s, there have been several major waves of technologies,
paradigms and best practices for building and running them. The first major
wave, characterized as *client/server computing*, emphasized the separation of
user interface, system dialogue application logic and data persistence concerns.
This was driven by the emergence of user-friendly GUI technology as well as the

A. Hameurlain et al. (Eds.): TLDKS VII, LNCS 7720, pp. 128–169, 2012.

ever increasing need to deal with enterprise system landscapes and enterprise application integration. Key ingredients of this wave were transaction monitors [3] and message-oriented middleware (MOM) [4], for example. The second major wave, characterized as *distributed object computing*, emphasized the use of object-oriented abstractions in the construction of distributed systems. Key examples of technologies supporting this approach include CORBA [5] and J2EE [6]. The third major wave, which is arguably still ongoing, is the *Service-Oriented Architecture* (SOA) wave which focuses on the Internet as the underlying distribution platform. Key examples of technologies in this wave are Web Services and the emerging REST (Representational State Transfer) services paradigm [7]. Finally, the fourth major wave that is just starting to emerge is *cloud computing* [8]. This emphasizes the flexible and site-transparent distribution of computing platforms and applications over outsourced data center infrastructures. Key examples of technologies in this wave are the Google Cloud, the Microsoft Cloud, Amazon Web Services and the on-demand data centers of companies like IBM and HP.

Whilst this rapid evolution of distributed computing technologies has provided a rich set of platforms and paradigms for building robust enterprise systems, it has also left a legacy of unresolved problems. The first is a trail of confusion and inconsistency in the concepts used to build distributed systems. Even within individual paradigms there are inconsistent interpretations of some of the most basic concepts, such as whether services are (or should be) stateless, for example. Furthermore, between the different paradigms there is little consensus about the core ingredients of distributed systems, for example, what are the differences and relationships between components, services and objects. The second problem is that the evolution of the different distributed system technologies has been overwhelmingly driven by the server-side concerns of the client/server divide rather than the client side. As a result, developers of regular client applications, business processes or mobile "apps" (e.g. Android) typically have to access server-side assets via low-level, platform-specific abstractions optimized for solving server-side problems rather than via those customized for their needs. This not only increases the effort involved in developing and maintaining client applications, it also increases its proneness to errors.

The importance of addressing this problem has grown as client applications have become more visible and have started to play a major role in the perceived usability of service-based systems. Moreover, in large enterprise system landscapes [9] services are often clients of each other, and experience has demonstrated that the most successful enterprises have paid particularly attention to the infrastructure features needed to support flexible service usage [10,11]. To date, however, SOA best practices have primarily focused on only one aspect of usability – the rapid and straightforward integration of arbitrary technologies (including legacy systems) and platforms at the server side of the client/server divide. However, this ease-of integration at the server side has come at the price of lower flexibility and ease-of-use of distributed software assets at the client side in client application development.

To address the basic problems and to reduce the artificial complexity involved in building client applications a unified conceptual model of service-oriented computing that supports both the needs of client developers and service providers is required. Our premise is that such a conceptual model should be completely independent of, and implementable on top of, the different distributed computing paradigms discussed before since these represent ultimately just implementation alternatives. Not only that, it should also be compatible with the implementation technologies and modeling languages used to realize client applications today, such as high-level programming languages or business process modeling languages. Just as high-level programming languages are optimized for human programmers rather than for realization platforms, we believe that the features of such a conceptual model should be determined by what is best for client developers rather than by the idiosyncrasies of individual implementation platforms.

In the previous papers [1,2] we have outlined the core conceptual ingredients of a client-oriented model of distributed computing assets in an informal way using a motivating example. This paper integrates and builds on material in these two papers to provide a complete description of the proposed unified modeling framework. It presents a concrete set of metamodels for the different abstraction levels and views involved in a distributed computing landscape and positions these within a single unified conceptual framework. The paper also presents two new examples of how specific platforms and technologies are accommodated in the framework. The key difference between our framework and other more general conceptual models for distributed computing like RM-ODP, Service Component Architecture (SCA) [12], CORBA Component Model (CCM) [5] is that our conceptual framework:

- has been designed to accommodate mainstream platforms and paradigms as special cases, and thus includes a transcending set of concepts that can help to unify and systematize the field of distributed computing in general,
- provides a focus shift onto the needs of client developers rather than the traditional server-side concerns (e.g. persistence, interoperability, transactions, robustness, etc.) that dominate distributed computing platforms. For client developers, these manifest themselves as non-functional properties (e.g. performance, reliability, etc.) wrapped-up in service-level agreements.

The rest of the paper is structured as follows. In Sect. 2 we first review the emergence of the different strands of SOA in the past and discuss the impact of the service-oriented metaphor on the usability and re-usability of software components. Then, in Sect. 3 we discuss the foundations of the unified conceptual framework and discuss how its ingredients fit together in our approach in general. Section 4 first presents the Metamodel Architecture of the unified conceptual framework and then introduces the Core Metamodel to capture the transcending concepts and to act as the common supermodel for all of the other metamodels. In Sect. 5 we then present the metamodels for the platform-independent views of the service- and the client-oriented perspective on a distributed system to provide optimized abstractions to the involved stakeholders. This is followed by the description of five key PIM-level realization patterns that can be used to map client-side abstractions to server-side abstractions and vice versa. In Sect. 6 we then present concrete metamodels for a service- and a client-oriented realization platform respectively. Section 7 introduces an example application scenario to show how our approach and its models can be applied using particular realization platforms (Web Services for the service-oriented and Java for the client-oriented perspective). Section 8 then discusses related work and finally Sect. 9 concludes with some final remarks.

2   Review of the Service-Oriented Metaphor

The term service-oriented architecture dates back at least to 1996 where it was used by the Gartner Group to identify a pattern for establishing a hub-and-spoke architecture for improving enterprise application landscapes. More concerns were gradually integrated into the service-oriented paradigm until eventually the SOA community somehow agreed upon a certain set of characteristics that make a service-oriented architecture, so that in the last decade definitions like the following one have become wide-spread and commonly accepted:

> *A service is a coarse-grained, discoverable, and self-contained software entity that interacts with applications and other services through a loosely coupled, often asynchronous, message-based communication mode*[13]

The problem with such definitions is that they merge rationales from different software engineering sub-disciplines that have picked up the service-oriented metaphor and use the service-oriented terminology in distinct ways. These sub-disciplines include:

- enterprise application integration,
- business-to-business integration,
- business process management and
- software productizing also known as SOA governance.

Figure 1 therefore also provides an overview of these different strands of service-oriented architecture. These are only the mainstream perceptions of SOA, which are all related to enterprise computing issues. There are other initiatives loosely related to the enterprise computing domain like OSGi and Jini which could also be considered as part of the landscape of SOA technologies. In concrete SOA projects there is often confusion about which view of service-orientation is actually relevant in a given situation. Moreover, the various design principles discussed for SOA are usually driven by one or the other motivation.

We are interested in SOA from the viewpoint of distributed object computing, because, given the trend in cloud-computing in general [14] and the foreseeable wave of cloud-based software engineering [15], the question of good design of distributed enterprise applications and enterprise application landscapes will continue to be a major issue in the future.

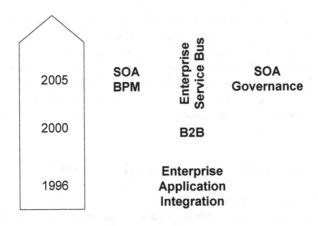

Fig. 1. Emergence of SOA disciplines

In the next section we first review the various strands of service-oriented architecture in order to set the stage for our discussion. Then, another section discusses the central aspect of usability and re-usability of services and how it is addressed by the basic SOA design principles.

2.1   Strands of Service-Oriented Architecture

SOA is not a single paradigm, and over time several visions and concepts have emerged. As mentioned previously in [6,7], initially, the service-oriented architecture emerged as a three-tier architecture for enterprise application integration. This happened shortly after Phil Bernstein envisioned a trend in which enterprise computing facilities would eventually become a kind of utility in enterprise system landscapes [4].

Shortly after the crash of the B2C hype in 2000, the B2C vision emerged. This was about creating a new lightweight, ubiquitous and open electronic data interchange platform based on concrete Web Service technologies, actually an alternative to the established EDI (electronic data interchange) technology [18]. As a result B2C proponents hyped the service-oriented metaphor, and Web Service technology has become a synonym for SOA for many stakeholders in the enterprise computing community. This is also the reason why SOA is today recognized as an enabler for flexible, adaptive business processes, simply because of the wide-spread usage of Web Service technologies in state-of-the-art commercial business process management (BPM) products. Another strand of argumentation in service-oriented architecture is in the area of software development methodology. Here, inspired by the new tools for component maintenance, and in particular those that deal with service discoverability, innovators where thinking of new opportunities to accelerate the software development process. For example, the systematic and massive reuse of software components in software development projects was envisioned. Recently, the massive reuse of software components stemming from the multitude of projects in a large enterprise is typically considered in the realm of so-called SOA governance [19]. In the extreme, we have even advocated this beyond organizational boundaries to view the World Wide Web as a service repository [20,21].

At the same time an intensified interest in SOA-based business process management technology and SOA governance lead the SOA community to introduce the notion of the Enterprise Service Bus (ESB). This returned the SOA discussion to its roots as an overall architecture for enterprise application integration (EAI).

2.2   On Service-Oriented Architecture for Reuse

One major goal of service-orientation is to optimize the usability of services. We explicitly mention usability and not re-usability in the previous sentence. In practice there has been lot of confusion about best practices related to the statelessness of services, because often, the conceptual level of discussion is not clearly separated from the implementation level. With respect to the implementation level the question of state refers only to the decision about where to maintain state, i.e., on the service hosting application server, or somewhere else, typically persistently in the database tier. However, from an architectural perspective the concept of statelessness actually stands, in a subtle manner, for a characteristic programming language design of the services as they are presented to potential clients. There is a direct tradeoff between the feature-richness of the implementation technology of the service interface and its immediate usability for client developers.

The less one assumes about the client and service programming language concepts and features, the less complex is the mapping needed between the concepts of a client programming language and the concepts of the service interface specification. In this paper, a mapping is not only a conceptual mapping between features, but the whole realization technology involved in bridging between the service interface and the client code. For example, the IDL compilation and generation of stubs in the CORBA world belongs to the notion of mapping. In the following, this means the less you assume about the client programming language concepts and features, the easier it is for a client to exploit the service. This results in an overall decoupling, but also in a loss of features and benefits of proven programming language constructs. In other words, SOA minimizes the infrastructural effort involved in using a service at the cost of increasing programming efforts.

A service interface or a service facade, which consists of several interfaces, can realize complex concepts like the construction and manipulation of objects. But the service facade itself is not complex. It is like a remote control unit with knobs that support operations on the complex concepts behind the facade. For example, a service can support object-based programming abstractions, but the service can also be used by a programming language that is not even object-based, e.g., PASCAL, FORTRAN or even assembler. All the client program has to do is to stimulate the service as required by the service's contract. Even the realizing service code does not necessarily have to be written in an object-based language. The structure of the overall system is object-based, because it results in the conceptual realization of objects. This reflects the fact that object-based and even object-oriented structures (simulating subtype polymorphism) can be programmed with a plain 3GL programming language. This of course is hard to do, because there is no support by special programming language features and tools (e.g. type checkers), but it nevertheless is possible.

All this is related to the complexity of the type system. To explain this we turn to an extreme case which we call the *totally flat services style*. Assume that we allow only basic types in the definition of service interfaces. Nevertheless, classes with attributes can be realized. Take the following OO code as a specification (not as a program):

```
class A {
  b:B;
  a1:Int;
  a2:Int;
}

class B {
  b1:Int;
  b2:Int;
}
```

This specification can be realized by services providing the following interfaces:

```
buildA (return ID:String, bID:String, a1:Int, a2:Int);
setB(ID:String, bID:String);
getB(ID:String, return bID:String);
seta1(ID:String, a1:Int);
...
...
buildB (return ID:String, b1:Int, b2:Int);
...
```

Now, for example, it is possible to construct an object net consisting of two objects using the following client code:

```
buildB(bhandle,1,2);
buildA(ahandle,bhandle,3,4);
```

This is meant to have the same semantics as the following OO code:

```
B b = new B(1,2)
A a = new A(b,3,4);
```

The above client code can be written in a programming language that is not an OO language. It can even be written in a basic programming language that does not support complex types like records, arrays, structs, i.e., only basic types. In practice, however, it makes sense to actually use an OO language and this will be supported by a mapping, i.e., a re-construction of the objects with the features of the OO client programming language. But the burden of creating the mapping is essentially an overhead. In the SOA world, it does not have to be provided by the infrastructure and the service provider. Although the burden of creating the mapping is overhead, it can be argued that client programming with SOA is nevertheless easier because client programming involves more than coding. You must code the mapping, but you do not have to adhere to a pre-defined/contracted mapping mechanism. Adhering to a predefined/contracted mapping mechanism can be considered (i) heavyweight and (ii) restricting.

The *totally flat services style* is a thought experiment, because in practice there are mediating type systems like the .NET Common Type System that represents a common denominator. However, a mediating type system is already a compromise that deviates from the notion of pure service-orientation. Pure SOA is an ideal based on the specification of interfaces that can be used immediately by any existing programming language (an ideal that obviously cannot be fulfilled). SOA essentially tries to find the sweet spot between immediate and lightweight usability of a service and robustness of its usage.

The tradeoff is similar to that described by John K. Ousterhout in his seminal paper on Scripting technology [22]. In this paper he distinguishes between system programming languages and scripting languages. System programming languages are third-generation languages that are feature- and concept-rich, in particular with respect to full-fledged type systems and a notion of type safety. In Ousterhout's taxonomy, system programming languages serve the purpose of implementing the various applications of an enterprise application landscape, whereas scripting languages are the optimal choice for implementing the glue between the enterprise applications in EAI scenarios as these represent lightweight, un-typed languages. Obviously there is a trade-off between the lightweightness of a programming language and the scalability and maintainability of the crafted code artifacts.

3   Foundations of the Unified Conceptual Framework

The ultimate goal of the unified conceptual framework presented in this paper is to provide a modeling and development framework in which the different concerns and roles in distributed application development can be expressed and related to one another, and the overhead (or accidental complexity) involved in using (as opposed to developing) service infrastructures is minimized. The basic concerns used to structure the framework are therefore the level of abstraction at which a system is represented as well as the roles from which the software entities of a system are viewed. A further premise in the design of such a unified conceptual modeling and development framework is that the client developer role should be equally, if not more, influential in defining the concerns that need to be taken into account than the service provider role which has traditionally been dominant. This also includes the explicit support of the missing notion of object types and instances in service-oriented development approaches. This is important both for the modeling of client applications in a client-oriented style and for the modeling of (distributed) services in a service-oriented style alike. In this section we therefore discuss these issues as the basic foundations of the unified conceptual framework.

3.1   Abstraction Levels

The Model-Driven Architecture (MDA) popularized by the OMG [23] represents software systems using multiple levels of abstraction. The most abstract level at the top is the Computation-Independent Model (CIM) in which models of the business processes to be automated or the environment to be enhanced are described independently of the envisaged IT-based solution. On the next level, the Platform-Independent Model (PIM) describes the key ingredients and the behavior of the envisaged system independently of the idiosyncrasies of specific platforms. The Platform-Specific Model (PSM) describes the system in terms of the concepts supported by a specific platform, but not necessarily in a way that is directly executable. In other words, a PSM is still a model even though it is platform-specific. Following, the lowest level of abstraction in the MDA is the executable implementation which does not require any further manual transformations (based on the assumption that the appropriate compilers and virtual machines are available). These levels correspond roughly to the classic abstraction levels of model-driven software engineering. Our approach adopts the basic MDA abstraction levels and the associated terminology. However, in the unified conceptual framework we focus only on the two central levels (PIM and PSM) since this is where the key separation of concerns as well as the identification of common abstractions takes place.

3.2   Roles

With the evolution of the different technologies of distributed computing from simple client/server approaches to complex and heterogeneous service-oriented architectures the differences between the concerns of clients (i.e. service users) and those of services (i.e. their providers) have grown tremendously. Moreover, these differences are set to increase even further as distributed computing evolves towards cloud computing. Therefore, the two fundamental roles of concern in distributed system and client application development are the **service provider** and the **client developer**.

Figure 2 shows an Enterprise Service Infrastructure (ESI) that provides three different services, implemented and maintained by one service provider and used by different client applications of two distinct client developers. A fundamental premise of the approach is that there is a clear boundary to the enterprise system for each stakeholder. As depicted in Fig. 2, the client developer on the left (Client Developer A) actually belongs to the organization owning the ESI, and thus is represented as being inside the enterprise system boundary, while

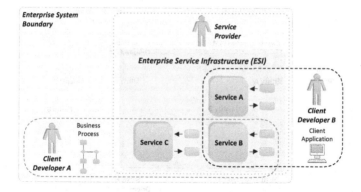

Fig. 2. Roles in Distributed Application Development

the other (Client Developer B) is outside the boundary and uses the services as a usual external customer. The clear distinction of the two different roles – **client developer** and **service provider** – and the consideration of their viewpoints and the boundaries is an essential prerequisite of our unified conceptual framework and the overall approach as this paper will discuss.

3.3   Two-Dimensional Modeling Space

Based on the distinction between the abstraction level (PIM/PSM) and the different roles and their viewpoints as discussed in this section, the overall structure of the unified conceptual framework is derived by regarding these dichotomies as orthogonal dimensions. The basic goal is to address and support the viewpoints of both roles on both levels of abstraction. As illustrated in Fig. 3, and as will be presented in this paper, for our unified conceptual framework these concerns result in a two-dimensional modeling space that consists of four different kinds of models. Their purpose will be explained in more detail in the following.

Service-Oriented PIM (SPIM). The role of a SPIM is to provide a platform-independent view of a service landscape from the point of view of a service provider. There is therefore no notion of, or support for, clients (i.e. client types) required in the SPIM. The abstractions used in a SPIM are independent of any particular realization technology as we will discuss in more detail later when we present the SPIM Metamodel. If numerous service providers are involved in supporting different parts of a single overall service landscape, they will each have a tailored SPIM that represents their own particular view of the landscape as depicted in Fig. 3.

As depicted in Fig. 3 the service-oriented view must not consist of one model that contains all entities as they are defined in the client-oriented view for a single client application scenario. The models may be distributed between different organizational units which are not related to each other. In case there is only one organizational unit providing all of the required entities there should be one model only containing all entities.

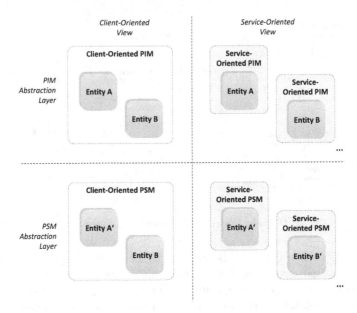

Fig. 3. Space of Models in the Unified Conceptual Framework

Client-Oriented PIM (CPIM). In contrast to the SPIM, a CPIM provides a platform-independent view (i.e. the abstractions) of one or more services, provided by one or more service landscape(s) tailored to the needs of a particular client developer and his client application scenario. A CPIM therefore includes the notion of, and support for, client types that may be related to one or more entities of the CPIM as we will introduce in more detail in Sect. 5.2. As with a SPIM, a CPIM transcends particular realization technologies and represents services as platform-independent client-oriented abstractions.

Service-Oriented PSM (SPSM). A SPSM provides a platform-specific representation of a service landscape (or a part of a landscape in case of multiple service providers) from the point of view of a single service provider. In terms of the traditional MDA paradigm it extends and refines the SPIM using platform-specific abstractions. The abstractions representing SPSM entities are usually refined towards service-oriented abstractions as expected by most of the realization platforms for distributed computing.

Client-Oriented PSM (CPSM). The CPSM in the unified conceptual framework represents a refinement of a platform-independent model (CPIM) adding detail through platform-specific concepts and abstractions. Ideally, this platform-specific detail will only relate to the client-side because the server-side abstractions (e.g. services) should be accessible as defined by the PIM-level abstractions. In fact, this means that the abstractions defined on the PIM-level are usually not radically changed, i.e., objects of the CPSM refine objects of the CPIM and they do not alter or derive directly from SPIM objects or even SPIM objects.

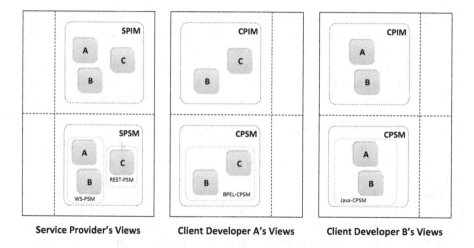

Fig. 4. Role-dependent relevance of Models

In general, not all models of this two-dimensional modeling space are of interest to all roles in the unified conceptual framework. Each stakeholder has a particular constellation of views that reflects his particular concerns as illustrated in Fig. 4.

Referring to the small example scenario that we have presented in Fig. 2, the service provider responsible for the whole ESI usually is only interested in the server-side abstractions of the ESI's services. Thus, these are all part of this service provider's SPIM and SPSM views. As also depicted in Fig. 4, single entities of the SPIM may be realized using different platforms in the SPSM. Which platforms are chosen therefore, depends on the requirements of single services of the ESI (e.g. on non-functional properties, existing infrastructures, etc.). Client developer A on the other hand, is only interested in his own particular client-oriented view of the ESI. This is reflected in his CPIM and CPSM which only contain abstractions of the service landscape's entities used by his client application scenario – in the example these are the abstractions of the entities B and C. Similarly, client developer B defines his own individual client-oriented views of the used entities tailored to his concerns. The CPIM and CPSM of client developer B therefore only contain abstractions of the services A and B. Although both client developers do not actively work on defining SPIMs or SPSMs, they at least need to use them to gather the knowledge about provided abstractions and the domain to define their CPIMs.

3.4 Support of Types and Instances

In current service-oriented development platforms, the notion of types and instances is generally not supported. Service-oriented platforms such as Web Services for example provide no explicit support for the notion of types. Services are usually viewed and treated as instances having their own unique identity and properties. Moreover, for the parameters exchanged between services and their clients there exists no notion of data abstraction. The internal representation of complex parameters is always explicitly laid out in the messages used for operation invocations and their responses.

While this emphasis on services and explicit revelation of data structures may be advantageous from the perspective of the service provider's role since it simplifies service integration and deployment, it is not necessarily advantageous from a single client developer's point of view – see Sect. 2.2, where we discussed the notion of the sweet spot between language design for service-intensive systems implementation in comparison to the usability and usage of services by client programs. The premise of the unified conceptual modeling and development framework is therefore to offer client developers the possibility to start on the PIM-level with familiar concepts and abstractions like object types, rather than forcing them to deal with the idiosyncrasies of service-implementation technologies on the PSM-level first. Our unified conceptual framework therefore provides client-friendly support for modeling and developing client application scenarios based on client types and types (i.e. abstractions) of software entities on both abstraction levels, the PIM and the PSM.

3.5   Perspectives of State

Applying the different viewpoints supported in the unified conceptual framework, we are able to mitigate another weakness of current approaches of service-oriented development related to the overall confusion about the statefulness of services as already introduced briefly in Sect. 2.1. In general, today it is considered a best-practice to implement services as stateless software entities in service-oriented development. However, this is only a useful best practice for integrating and managing functionality on the server-side, especially when its primary role is to provide access to databases. This model (based on function-oriented, stateless abstractions) is not optimal from a client developer's perspective for the integration of services into client applications.

In fact, from a client-oriented perspective the overwhelming majority of services appear to be stateful, even if the implementing service provider would characterize them as stateless. In our framework we therefore distinguish between two essentially different kinds of state for software entities – the **algorithmic** and the **cohesive state** as defined in [2]. The algorithmic state refers to processes (i.e. functions) and the state of their execution (i.e. the internal progress of their execution), while the latter, the cohesive state refers to the state (i.e. the internal state of the abstraction as a whole) that spans multiple operations that belong to a cohesive abstraction.

In service-oriented development there exist both, functional abstractions (i.e. processes) and cohesive abstractions (i.e. objects) offering multiple operations alike. However, the ever ongoing discussion of statefulness clearly relates to the cohesive state of objects. To solve this dilemma, the unified conceptual framework introduces a novel distinction of cohesive state into two different kinds of state – the **observable** (cohesive) state of an object and the **inherent** (cohesive) state of the object as defined precisely in [2].

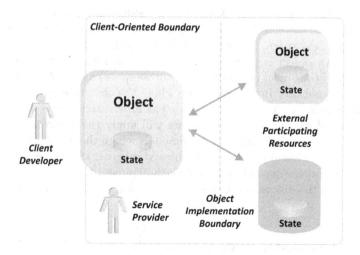

Fig. 5. Observable and Inherent State of Objects

As depicted in Fig. 5, a stateful abstraction of an object (i.e. observably stateful from the perspective of a client developer outside of the client-oriented boundary) can in general be realized in two different ways. Either the implementation of the object itself maintains the state internally (i.e. within the process boundary of its environment) or the state is deferred to external resources (e.g. a database or another object). But still, if the implementation of the object itself is stateless and the state is deferred to external resources, its abstraction visible to a client developer can be regarded as observably stateful (e.g. by elimination/encapsulation of the session ID parameters in the objects operations). We consider this kind of state as the **observable state** of an object. In general an object is regarded as **observably stateful** when it has observable dependencies for at least one of its offered operations, i.e. either an operation depends on an external participating resource or it depends on the object's internal state that can also be affected by other operation invocations. In the case that there exist no dependencies between the operations of an object, i.e. its operations always return the same values for different, arbitrary invocation sequences and call parameters, an object is regarded as **observably stateless**.

The other kind of state is referred to as the **inherent state** of an object. While observable state is relevant at a conceptual level, inherent state is a service implementation issue. Much confusion in the discussion of statelessness of services stems from the confusion between the conceptual and the implementation level. Observable state is considered at the PIM level whereas inherent state occurs on the PSM level in concrete platform-specific models. For inherent state the perspective on the object needs to be limited to the object's implementation boundary defined in general by the local run-time process space the object instance will be executed in. More specifically, this means that only contained objects are regarded as being inside the implementation boundary. Used objects therefore represent software entities that are either provided by another process space, i.e. other external objects (services), or a database for example. An object is therefore regarded as inherently stateful only if the internal state of the object (i.e. within its process space) has effects on the return values of its operations. In the other case where the observable state is deferred to external resources, the object is regarded as inherently stateless.

In the unified conceptual framework we will apply the different concepts of state to the entities of the different metamodels where these are relevant. With the concrete distinction between observableand inherent statefulness, the explanation of their mutual dependencies and the assignment to concrete metamodels we substantiate the discussion on service statelessness that we have outlined in Sect. 2.

4   The Metamodel Architecture

In the previous section we have outlined the space of models of the unified conceptual framework. The different models are not independent, of course, but are related to one another in carefully defined ways. In this section we therefore first present a metamodel architecture to organize them in a structured way. We then present the Core Metamodel (CM) as the root of the metamodel hierarchy which captures the commonalities of the different metamodels. In fact, one of the main contributions of our approach is the identification of the commonalities in the Core Metamodel and the capturing of the differences in the different client- and service-oriented PIM and PSM metamodels.

4.1   Structure of the Metamodel Architecture

For the unified conceptual framework we define the potential set of concepts and abstractions within one metamodel for each combination of the different views and abstraction levels. The resulting set of metamodels is organized within a specialization hierarchy of metamodels as presented in Fig. 6. At the root of this hierarchy the Core Metamodel defines the basic abstractions that are common to all roles and abstractions levels in client- and service-oriented computing alike. All of the other metamodels introduced in the metamodel architecture therefore inherit directly or indirectly from the Core Metamodel which therefore provides the corner-stone to implement the unified conceptual framework's metamodel architecture.

As depicted in Fig. 6, the metamodels for the PIM abstraction level – the CPIM and the SPIM Metamodel are direct specializations of the Core Metamodel. On the next level of the specialization hierarchy, the CPSM and SPSM metamodels for specific realization platforms are specializations of the CPIM, respectively the SPIM metamodel.

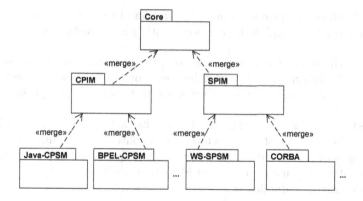

Fig. 6. Package Structure of the Metamodel Architecture

As indicated in Fig. 6, the unified conceptual framework may be extended by PSM metamodels for specific platforms for client- and service-oriented development alike, i.e. they may extend CPIM or SPIM metamodels. The prerequisite therefore is that the concepts to be defined in a certain PSM metamodel are already defined or represent extensions of already defined concepts of the PIM metamodel they inherit from.

4.2   The Core Metamodel

The Core Metamodel introduced in this section supports the three basic ingredients of imperative computing – *processes, objects* and *data types.* These are arranged within a single hierarchy in the metamodel as abstract classes that extend the basic abstract class *Entity.* They are used as the basis for further concepts as depicted below in Fig. 7.

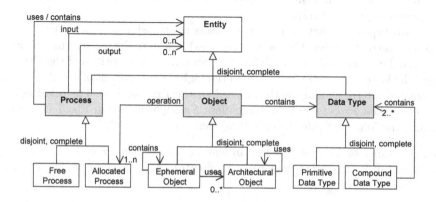

Fig. 7. Core Metamodel

In the following we provide a brief description of each class contained in the Core Metamodel and explain its relation to other contained classes.

Entity. At the top of the Core Metamodel, the abstract class *Entity* serves as the root of the inheritance hierarchy. Instances of subclasses of the class *Entity* therefore may represent any kind of software entity within a computing system.

Data Type. The class *DataType* is also an abstract class that essentially represents a value or a set of values whose use is controlled by a number of rules. The instances of its two subclasses *PrimitiveDataType* and *CompoundDataType* represent values that have no inherent identity of their own and they cannot be created, destroyed or duplicated.

Primitive Data Type. As a subclass of *DataType*, this class is used to represent the classic primitive data types such as *Integer, Character, String,* etc.

Compound Data Type. The subclass *CompoundDataType* is used to represent data types whose values are composed of combinations of primitive and/or compound data types, similar to records or structs as used in early days programming languages like Pascal or C. An entity of this type essentially contains at least 2 other data types, either both of the kind *PrimitiveDataType*, both of the kind *CompoundDataType* or mixed.

Process. The abstract class *Process* represents a functional abstraction that internally is comprised of a set of steps arranged in some well-defined order to achieve some effect or reach a certain goal. The *Process* type encompasses various programming-level abstractions such as subroutines, functions, procedures and methods, as well as high-level notions of processes such as workflows and business processes. Processes can have input and output parameters which can be instances of the entity *Object* or *DataType*. Since processes involve the sequential execution of sub-steps to achieve a goal, processes have an associated notion of **algorithmic state** that represents their current progress through the designated sequence of steps.

Process types may have a *uses* or *contains* relationship to other entities contained in the metamodel. The former (i.e. *uses*) relates to the use (the invocation) of external entities that are not defined as internal, owned entities of the *Process* type. The latter (i.e. *contains*) refers to the logical containment of entities which are created internally and are thus owned by the *Process* type. The relation between the class *Process* and the basic class *Entity* as drawn in Fig. 7 therefore requires the following rules (restrictions) defined:

- an instance of *Process* may *use* an instance of another *Process*
- an instance of *Process* may *contain* (create) an instance of an *Object* or a *DataType*

The abstract class *Process* finally is partitioned into two different sub-types defined by its sub-classes *FreeProcess* and *AllocatedProcess*.

Free Process. The class *FreeProcess* is used to represent process types that are self-contained and are not allocated to any particular object. It therefore represents a purely functional abstraction (akin to a function, procedure or subroutine in older programming languages). Typically, business process templates (e.g. BPEL) can be considered free processes that do not belong to any cohesive abstraction (e.g. an object). Free processes that occur at the service side are mere computational services. Also, information services, like weather forecast and stock exchange Web Services should be considered as free processes. Information services can and should vary over time, i.e., in a sense they are also state-dependent. But the state they depend on is beyond the control of the client. Therefore, pure information services can be regarded as examples of free processes.

Allocated Process. The type *AllocatedProcess* is a class of the metamodel used to define a process that belongs to an *Object*. The existence of an instance of an *AllocatedProcess* is therefore tied to the existence of the *Object* that it belongs to and it cannot be used by other entities independently of that object. Allocated processes have direct access to the cohesive state of the object they logically belong to and therefore correspond to methods in object-oriented terminology.

Considered from a service-oriented perspective, allocated processes are the standard for the important class of business-related services, e.g. services that realize Enterprise Resource Planning (ERP) functionality, since basically, the purpose of these is the multi-tenant assessment and maintenance of business objects and means of reporting on business objects. The information and maintenance of business objects clearly falls into the realm of allocated processes.

Object. An *Object* class is used to represent the basic object abstraction familiar from object-oriented and object-based computing in general. The basic characteristic of an object is that it encapsulates one or more attributes of a *DataType* behind one or more *AllocatedProcess* entities, i.e. methods. *Object* instances have their own unique identity and unify the core ingredients of function-oriented computing – processes and data types. The key additional contribution in the unified conceptual framework is the definition of two basic kinds of *Objects* – ephemeral and architectural objects. The class *Object* is therefore an abstract class since it has no direct instances. Because they encapsulate data values, objects have an associated notion of **cohesive state**.

Architectural Object (AO). The class *ArchitecturalObject* is used to represent a stable object whose instances have a lifetime that normally coincides with the lifetime of the system as a whole, because they are the *components* (i.e. indispensable parts) of the system. The only exception is when architectural reconfigurations are performed. Architectural objects are in general large, behavior-rich objects and are usually few in number. In fact, most of the time there is only one instance per *ArchitecturalObject* type in a system. The different notions of components, services and distributed objects as found in contemporary distributed computing technologies and literature are in general encompassed by the notion of architectural objects.

Ephemeral Object (EO). In contrast, the class *EphemeralObject* represents a sub-class of *Object* whose instances usually have a temporary lifetime with respect to the lifetime of the complete system, (i.e. relative to their owning system). In other words, instances of this kind are frequently generated and deleted during the lifetime of a system and essentially correspond to data entities. They correspond to concepts such as *Data Transfer Objects* (DTOs) and *Entity Beans* in J2EE [6] in specific realization technologies.

With the definition of these classes the Core Metamodel provides the foundation for the unified conceptual framework with clearly defined and characterized entities that go beyond the usual applied terminology in the different well-known software development and engineering paradigms.

5   PIM Metamodels

The next level in the metamodel architecture is the PIM-level. In this section, we consider the two metamodels at this level within the overall metamodel architecture. Both the SPIM and the CPIM metamodel inherit from the Core Metamodel and extend its concepts and abstractions by additional platform-independent concepts optimized towards the service-provision and the client-oriented perspectives. The following sections provide a detailed overview of these metamodels and explain their extensions. Finally, we introduce some basic PIM-level realization patterns that may be used to map an abstraction of one perspective to an abstraction of another perspective and vice versa.

5.1   Service-Oriented PIM Metamodel

The SPIM Metamodel, depicted in Fig. 8, mainly provides extensions of the entities *EphemeralObject* and *ArchitecturalObject*. In the following paragraphs we discuss these and introduce concepts that directly relate to the state behaviour of abstractions.

Ephemeral Object Extensions. The class *Ephemera Object* of the SPIM Metamodel is divided into two subclasses, *DataObject* and *Behavior-RichObject*. The former subclass represents a type that is purely a wrapper for sets of attributes. In other words, data objects essentially wrap up and encapsulate the data contained in an instance of a *CompoundDataType* by shielding the attributes from direct access via setter- and getter-operations. By definition, *Data Objects* therefore may only contain Create, Read, Update and Delete (CRUD) operations or attribute-related setter- and getter-operations and they provide no rich behavior. This is what basically distinguishes a *DataObject* type from a *Behavior-RichObject* type. The latter provides extra functionality by allocated processes (i.e. methods) that calculate new information that is not directly stored in the attributes.

Both types of *Ephemeral Objects* are always stateful abstractions as they encapsulate attributes that represent their internal state explicitly or implicitly.

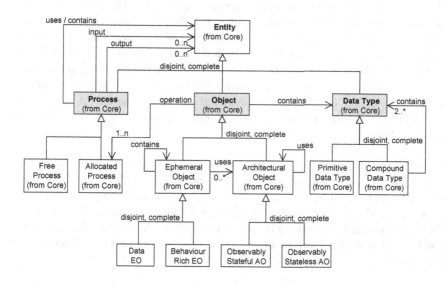

Fig. 8. SPIM Metamodel

Architectural Object Extensions. The *Architectura Object* class taken from the Core Metamodel is also divided into two further subclasses – *ObservablyStatefulArchitecturalObject* and *ObservablyStatelessArchitecturalObject*. These capture the distinction between observably stateful and observably stateless objects that we discussed in Sect. 3.5 and [2]. *ObservablyStatefulArchitecturalObjects* are objects that represent a stateful abstraction (i.e. they have cohesive state) from the point of view of a client. In contrast, *ObservablyStatelessArchitecturalObjects* have no observable cohesive state and are thus always regarded as stateless abstractions.

5.2   Client-Oriented PIM Metamodel

The CPIM Metamodel provides an extension of the Core Metamodel that includes the general, platform-independent concepts used to support the modeling of a client application scenario using the services provided by a distributed system. Like the SPIM Metamodel, the main extensions are again applied to the classes *EphemeralObject* and *ArchitecturalObject*, but the *Process* class is also extended to support the notion of client types in the CPIM Metamodel. Additionally, the CPIM Metamodel extends the different concepts into a form that is more natural for client developer when modeling client applications. These properties, that we introduced in [2], essentially reflect the behaviour of a type's instances at run-time as well as the static relations of a type to client types. Thus, we first clarify the notion of client types before we present the different classes of the CPIM Metamodel.

As one of the basic requirements of our approach is the ability to model client application scenarios that use entities of a distributed system in a more client-friendly way, it is not only necessary to define the required types of software entities, but also to define the different client types of an application scenario that use and access these. A client type therefore is defined as a self-contained entity whose instances are used to provide (parts of) the client application scenario's desired functionality or purpose. As depicted in Fig. 9, the CPIM Metamodel defines different classes to represent the two kinds of client types – *ClientProcess* and *ClientApplication*.

Client Process. The class *ClientProcess* in the CPIM Metamodel represents an extension of the class *FreeProcess*. Instances of this class represent self-contained clients with a clearly defined functional interface. The execution of a client process is triggered once, externally, by the invocation of the provided interface. A *ClientProcess* instance may only *use* (i.e. invoke) other instances of the class *ClientProcess* and may *contain* instances of type *Object* and *DataType* like its parent class.

Client Application. The other client type specified in the CPIM Metamodel is defined by the class *ClientApplication*. Compared to a client process, instances of this type of client are started usually by human interaction using an application's entry point. Client applications usually provide a (rich) set of functionality that is defined by the use cases related to the application. The execution order of the provided functionality (i.e. of the different use cases) is usually driven arbitrarily by the human user and is restricted only by pre-defined rules. In the context of the CPIM Metamodel, the *ClientApplication* type is defined as an extension of the class *EphemeralObject* since instances of it have an identity, a cohesive state and a lifetime that is often much lower than the lifetime of the system they are cooperating with. In terms of relations to other entities within the metamodel, the class *ClientApplication* is subject to the same rules as its parent class.

The remaining set of extensions provided by the CPIM Metamodel in principle corresponds to the entities defined in the SPIM Metamodel. However, as we mentioned in the introduction to this section, the difference is the assignment of different client-oriented properties to characterize them from the client-oriented perspective. In the following we therefore provide a brief discussion of the client-oriented properties *dynamic, static, private* and *shared* and their application to the entities in the CPIM rather than repeating the definition of already introduced classes.

Dynamic vs. Static Object Types. One of the most important properties of a type from the client-oriented perspective is whether it is dynamically instantiable at run-time or whether a fixed number of instances are needed at run-time. This is because objects in software applications typically play either the role of an architectural object or the role of an ephemeral object as we specified in the Core Metamodel. A client developer therefore may choose whether a type represents a **static** type needed at run-time with a limited, pre-defined

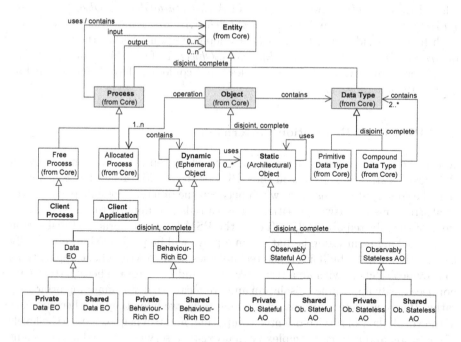

Fig. 9. CPIM Metamodel

and fixed number of instances (from the perspective of the client) or as a type where the number of instances may vary at run-time as the client can generate instances on the fly. The latter are therefore referred to as **dynamic**. Instances of *static* types are usually provided by the environment, i.e. they are never instantiated explicitly by their clients. As shown in Fig. 9, *EphemeralObject* types are therefore characterized as *dynamic* types while *ArchitecturalObject* types are characterized as *static* types.

Private vs. Shared Object Types. The characterization of a type as *dynamic* or *static* is related to the point of view of a single, individual client type instance. Usually, however, there can be more than one instance of a given kind of client running at a particular point in time as defined by the application scenario. This is true whether the client is a process or a typical client application. Taking the distinction between client instances and types into account, an additional important dichotomy between the objects used by clients becomes apparent. This relates to the distinction between objects that are **private** to each individual client instance, or objects that are **shared** by client instances. This distinction is orthogonal to the previous distinction so that *dynamic, private* types as well as *dynamic, shared* types are possible.

The class *EphemeralObject* from the Core Metamodel is therefore further divided into two subclasses – *PrivateEphemeralObject* and *SharedEphemeralObject*.

The CPIM also defines two subclasses of *ArchitecturalObject* – *ObservablyStatefulArchitecturalObject* and *ObservablyStatelessArchitecturalObject* – which are both further divided into two subclasses depending on whether they are *shared* or *private* similarly to the class *EphemeralObject*. In fact, this is orthogonal to the property observably stateful/stateless, and could also have been modeled as a separate generalization set.

5.3   PIM-Level Realization Patterns

In this section we define realization patterns which represent mappings between PIM-level abstractions (either in the CPIM or the SPIM). These patterns represent the core strategies used to switch between the different possible perspectives of abstractions of software entities. They can also be applied to the specializations of the affected abstraction (i.e. at the PSM level) as a practical realization step. The long term vision is that the mappings will be performed automatically by (i.e. they will be built into) the technologies used to write client applications to provide the client with an alternative view, and vice versa. These patterns are not necessarily applicable in isolation and are meant to define mapping principles rather than complete solutions. In the long term, we hope they will become part of the vocabulary of service-oriented computing, and will help to reduce the levels of confusion and artificial complexity that exists in service-oriented development today. Each of the patterns as presented in the following can theoretically be applied in both directions. However the names of the patterns as presented in Fig. 10 reflect one particular direction that we discuss in this section. In the following we provide a brief discussion of each pattern.

Data Type Reification. This pattern maps a *DataType* to an *EphemeralObject*. It essentially *reifies* a pure *DataType* whose instances represent collections of values, into a *DataEphemeralObject* type (i.e. a class) whose instances represent objects that can store the corresponding values. But, these store the values as private attributes that can only be accessed by setter- and getter-operations. In non-technical terms, it is used to turn a naked record into an encapsulated object. This pattern can be used to provide an object-oriented view of the structures (e.g. XML documents) communicated by Web Services over the internet. This essentially corresponds to the idea of DTOs in J2EE [6].

Ephemeral Object Externalization. This pattern maps an *EphemeralObject* type to an *ArchitecturalObject* type. For a given type of the former it defines an equivalent type of the latter that supports the same operations, but with an additional parameter for the identifier of an instance of the *EphemeralObject* class. This extra information is required because a single instance of an *ArchitecturalObject* type is responsible for storing the state of all instances of the *EphemeralObject* type which need to be distinguishable. In a sense, therefore, this pattern delegates the responsibility for maintaining the cohesive state and functionality of an *EphemeralObject* type to an *ArchitecturalObject*.

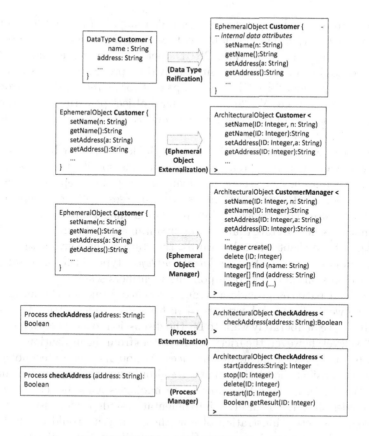

Fig. 10. PIM-Level Realization Patterns

Ephemeral Object Manager. This pattern also maps an *Ephemeral Object* type to an *Architectural Object* type. In fact, it is essentially an extension of the previous pattern. Additionally to the mapping of the operations of an *Ephemeral Object* type to the *Architectural Object*, including the extra identifier, its parameters and/or return values it also adds creation and deletion operations to complete the CRUD functionality. Furthermore it adds search operations and filters to find specific instances of the *Ephemeral Object* via their unique identifier or via its attributes.

Process Externalization. This pattern maps a *Process* type to an *Architectural Object* type. A single *Process* entity, with a given set of input parameters and/or result is mapped to an *ArchitecturalObject* type with a corresponding re-entrant (or idempotent) operation. Any client wishing to invoke the process therefore simply calls this operation on the singleton instance of this type that is created in a system. How the operation works is completely invisible to the clients of the process. In a sense, therefore, this pattern delegates responsibility for the functionality and algorithmic state of multiple instances of a given *Process* type to a single instance of an *ArchitecturalObject* type.

Process Manager. This process pattern also maps a *Process* type to an *ArchitecturalObject* type. The difference is that, like the *Ephemeral Object Manager* pattern, this pattern provides CRUD operations to create, manipulate and destroy process instances. This requires individual process instances to be identified by a unique identifier (ID) that is passed to, or returned by, invocations of the CRUD operations. While the ephemeral object patterns handle the cohesive state of instances of the type *EphemeralObject*, process patterns handle the algorithmic state of *Processes*.

These patterns capture some of the strategic decisions that have to be made to realize service-oriented systems, regardless of the specific technology used to implement them. They are therefore strategic platform-independent decisions. It is also important to note that clients are processes or objects, depending on their exact nature. Thus, when the appropriate manager pattern is applied, the algorithmic or cohesive state that is being maintained by the architectural object is the state of the client – in other words, the state of the conversation between the client and an instance of an *ArchitecturalObject* type. Such conversations are often called sessions and the corresponding identifiers sessions IDs.

The defined patterns provide a strong reference frame for the mitigation of frictions between clients and services when typical implementation technologies are involved. We deliberately want to lift the discussion to a conceptual level (i.e., to the PIM level) however, the frictions often stem from the utilization of different styles of technologies for clients and services. Although concrete technologies are involved, we consider the frictions that arise as only superficially implementation-related. We do so, because the concrete technologies embody certain concepts and rationales that are themselves implementation-independent issues.

To provide a better illustration of how these patterns would be used, let us have a closer look at the important *Ephemeral Object Externalization* pattern and a concrete technological scenario. Assume that we have a tier of services that shows itself with the common Web Services standard WSDL and a client tier programmed in the Java programming language. In the WSDL type system and interface concept there is no support for objects in the style of object-oriented programming, (i.e., there is no concept of ephemeral object as discussed in Sect. 4.2). This fact is also reflected by the Web Services SPSM Metamodel that we define in Sect. 6.1. Let us assume that the Web Service tier nevertheless conceptually realizes ephemeral objects (e.g., because it realizes an ERP functionality). This should then also be reflected in the corresponding SPIM as a matter of good service documentation. The mapping between the ephemeral objects of the SPIM and the architectural objects of the SPSM then exactly follows the described *Ephemeral Object Externalization* pattern. Java client programs that utilize these services perfectly fit because Java as an object-oriented programming language supports the ephemeral object notion.

At the implementation level Java code must utilize the implementing Web Services. The usage of each single service call to an architectural object must be designed again along the lines of the *Ephemeral Object Externalization* pattern. If there are only a few calls to architectural service objects the client code could craft the resulting Java objects each time on the fly. If there are a lot of calls, a more systematic approach is desired. A solution is to implement two access layers at the client side. The first access layer would consist of mere Java language wrappers to the Web Services and the second layer would provide the corresponding ephemeral objects and implement the necessary translation as prescribed by the *Ephemeral Object Externalization* pattern behind the scenes. Such access layers form a natural candidate for generative programming [24]. Actually, one of our visions for SOA tools is to provide such generators that mediate at the PSM level between the implementing technologies, which can be characterized as a lightweight rebirth of the CORBA IDL compiler approach in the more flexible setting of SOA.

6 Concrete PSM Metamodels

In this section we provide two concrete PSM metamodels for specific client- and service-oriented implementation technologies to be applied in the unified conceptual framework. We therefore consider the most widely known service realization technology – Web Services, and one of the most widely used client realization technologies – Java.

6.1 Web Services SPSM Metamodel

The first generation of Web Service standards (WSDL and SOAP) and several WS-Extensions represent one of the most well-known and widely used service realization technologies. However, for the definition of the Web Services SPSM Metamodel not all of the concepts of the SPIM Metamodel are directly supported restricting the use of some of the SPIM concepts.

Also, one of the functions of an SPSM is to indicate where a SPIM abstraction has a different name in a specific realization technology, or in other words, where a concept of a specific platform directly corresponds to a concept of a SPIM abstraction. This is achieved by placing the platform-specific name in parentheses after its platform-independent name.

This renaming is most evident in the case of *DataType* entities. Web Services basically support both contained specializations of the kind *DataType* as XML data types, also shown in Fig. 11. This SPSM Metamodel also indicates that the type *ArchitecturalObject* is referred to as **Service** in the Web Services terminology. Furthermore it defines that an *AllocatedProcess* entity corresponds to a Web Service **Operation**. As contained in Fig. 11, the Web Service SPSM Metamodel only contains the class *ArchitecturalObject*. In terms of the Web Services technology model, a Web Service cannot be instantiated by a client and needs to be provided by the environment. Therefore it does not include the notion of *EphemeralObject*. However, both subtypes of the class *ArchitecturalObject* are supported and the class *ObservablyStatefulArchitecturalObject* is further divided into two subclasses – *InherentlyStatefulArchitecturalObject* and *InherentlyStatelessArchitecturalObject*. This distinction reflects a PSM-level concern

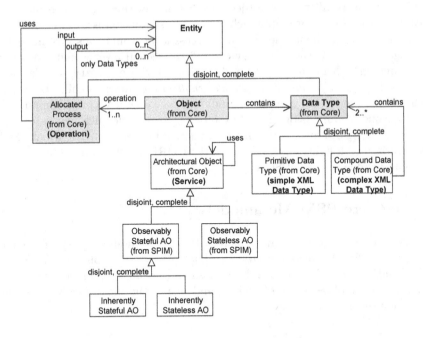

Fig. 11. Web Services SPSM Metamodel

and the fact that instances of the class *ArchitecturalObject* (i.e. Web Services) that appear to be stateful to clients (i.e. observably stateful) may not maintain any state directly on their own from a service-oriented point of view, but may delegate their logical internal state to a participating resource that is not visible to the client as discussed in Sect. 3.5. Such types are referred to as *InherentlyStatelessArchitecturalObject*. In contrast, an instance of the type *ObservablyStatefulArchitecturalObject* may indeed encapsulate the state that it exposes to clients internally. As discussed in [2] the failure to distinguish between these two properties (observable versus inherent statefulness) is the root of much of the confusion surrounding this issue in contemporary service-oriented technologies. By providing an explicit model of the distinction and relationship between these two concepts and allowing architectural objects to be characterized accordingly, the state-related behavior of services can be much more clearly understood by client developers and service providers alike.

Another restriction to the SPIM relates to the distinction between the classes *AllocatedProcess* and *FreeProcess* of the SPIM Metamodel. The Web Service standard does not recognize the concept of free processes and requires all processes to be allocated to objects (i.e. an operation is always allocated to a service and cannot exist or be provided remotely on its own). As contained in the SPSM metamodel, an operation of a service defines a *uses* relationship to the entity *ArchitecturalObject* (i.e. service) and it may invoke other services (i.e. the operations of other services). Furthermore there are restrictions defined in the *input* and *output* relations of operations as these accept only instances of the class *DataType* as parameters in terms of the Web Services technology model.

6.2   Java CPSM Metamodel

To illustrate an example of a CPSM Metamodel for the unified conceptual framework, we have chosen the Java platform that offers one of the most widely used programming languages for writing client applications. As a mainstream object-oriented language for regular GUI-driven programs that run on laptops and PCs and more recently in terms of applications for the rapidly expanding Android market, Java is able to support the concepts of the CPIM Metamodel in a fairly direct way as presented in Fig. 12.

The only entities from the CPIM Metamodel that the Java CPSM Metamodel does not support are given by the classes *CompoundDataType* and *FreeProcess* since data composition and behavioral abstractions are achieved using objects. Therefore, it only contains the process type *AllocatedProcess* and the data type *PrimitiveDataType* on the second level of the hierarchy.

Furthermore the client-oriented properties that have been an essential part of the CPIM Metamodel are removed in the Java CPSM Metamodel. This is in general true for all CPSM metamodels as we expect the generation of platform-specific software artifacts using the information supplied on the CPIM level.

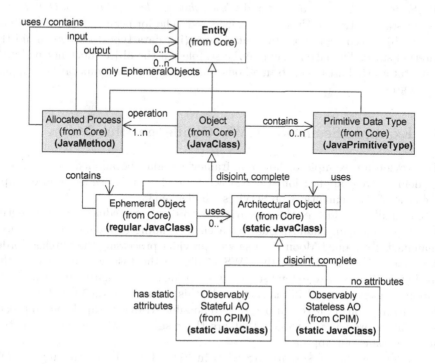

Fig. 12. Java CPSM Metamodel

In practice, this requires technical support by the framework that is aware of all model specifications and the transformation patterns and rules to be applied for the software artifact creation. In the step of implementing the client application it is only required to embed the respective artifacts that already technically should support the characteristics of the entities as specified on the PIM abstraction level.

Referring to the Java concepts and the common terminology used in Java environments the relation of the concepts of the Java CPSM Metamodel and the concepts of Java is defined as follows. Basically an *Object* type in the Java CPSM metamodel as presented in Fig. 12 corresponds to the concept of a Java *Class*. These may have one or more processes of type *AllocatedProcess* assigned that correspond to the concept of Java *methods*. Finally, the class *Primitive-DataType* corresponds to the concept of *primitive types* in Java. The two sub-classes *EphemeralObject* and *ArchitecturalObject* can also be related directly to concepts of Java. According to the definition of the former in the CPIM Meta-model, they correspond to *regular classes* that can be considered as *dynamic* as they can be instantiated arbitrarily at run-time as needed. In contrast, the latter need to be referred to the Java concept of *static classes* as we have characterized them as being provided by the environment. In Java instances of these types are provided statically by the Java Runtime Environment upon the start of the application (i.e. within the process of class loading upon an application start). Finally to characterize the subclasses of the type *ArchitecturalObject* – *Observably-StatefulArchitecturalObject* and *ObservablyStatelessArchitecturalObject* – the use of static attributes has to be considered. The former may be only stateful if the object type contains static attributes (i.e. data types) that maintain the internal state. The latter may only be stateless if the object does not maintain any static attributes (i.e. its methods only make use of temporarily available attributes).

7   Example Scenario – SWS Challenge Blue-Moon

To provide an example of how our framework may be applied in practice in modeling a real application scenario, we have chosen the Blue-Moon example scenario of the Semantic Web Services Challenge [33].

Originally, the idea of this scenario was to create a mediator software entity between a company named Blue (as a client sending a purchase order) and a manufacturer named Moon (as a service provider processing the purchase order as depicted in Fig. 13). In the SWS challenge the task was to develop this mediator entity and to extend service descriptions to automate the mediation to Moon's legacy backend infrastructure. In this section, we therefore first give a brief description of the scenario, discuss a possible strategy to apply our approach and finally provide the models that are the result of realizing the application scenario using our approach.

As depicted in the scenario overview in Fig. 13, the Blue company submits a purchase order to the manufacturer Moon whose mediator has to carry out several tasks (i.e. requesting different services of different backend systems) to go through the single steps required to process the purchase order. We adopt this example scenario and consider the mediator as a client type and Moon's legacy infrastructure within one client application scenario that focuses on the client-oriented design of the mediator. The services provided by Moon's legacy landscape therefore represent the distributed software entities to be used for the realization of the client application scenario. We simplify the last step of the depicted process (the call back initiated by Moon's *Order Management System* (OMS) to confirm order line items) and assume a delayed (e.g. 5 minutes) synchronous call to the OMS system to actively request an order confirmation after the order has been closed.

Basically, this scenario involves three roles which are the client developer of the mediator, the service provider of the *Customer Relationship Manager* (CRM) and the service provider of the OMS. Although both systems are within the enterprise system boundary of Moon, we assume they are maintained by different organizations. Furthermore this scenario dictates the strategy for realizing the client application scenario. This strategy obviously is driven by the already provided service infrastructure (i.e. Moon's legacy backend).

The starting point for this scenario therefore is defined by the two Web Services SPSMs including the provided distributed services by the Moon legacy system service providers. In general, for the creation of the CPIM, the client

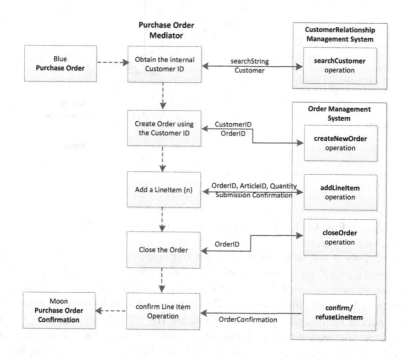

Fig. 13. BlueMoon SWS Challenge Scenario

developer at least requires some knowledge about the provided services (i.e. the CRM and the OMS). This knowledge is provided by the Web Service SPSM models, i.e. the PSM specification of the provided services. As indicated in Fig. 14, two potential ways can now be envisaged for the creation of a concrete CPIM of the mediator client application scenario.

The first approach (a) would be to reverse engineer the SPSM towards a SPIM provided to the client developer as a PIM-level specification. The other approach (b) directly provides the PSM-level specification (i.e. the SPSM) to the client developer. In both cases, the client developer has the opportunity to explore the provided abstractions and to make a decision about whether they fit his requirements or if he wants to use the gathered knowledge to design the CPIM using abstractions according to his preferred paradigm (e.g. using object-oriented abstractions elaborated in domain analysis and modeling) on the PIM-level.

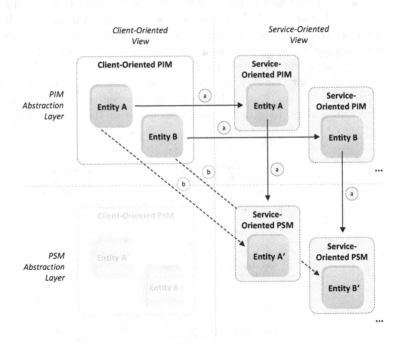

Fig. 14. Realization Strategy for the Mediator client

However, we want to show that using our approach the client model does not necessarily need to accommodate the abstractions provided by the SPSMs. Rather, it should be possible to create a client model (i.e. a CPIM) based on user-preferred abstractions. In the following sections we have chosen to follow the realization strategy (a) and first consider the SPSMs that deliver what is provided by the legacy backend infrastructure. We then reverse engineer the SPSMs to SPIMs and then present a CPIM of the mediator client application scenario based. Finally, we discuss how the defined CPIM abstractions can be supported on the PSM-level with the help of a human integrator and platform-specific stubs of the defined entities.

7.1 SPSMs of the Legacy Systems

For the Blue-Moon example scenario that we consider in this section, we omit the definition of SPIMs and assume that these have not been defined. Rather, we assume that the service providers of the CRM and OSM systems have started modeling their services on the PSM level as they planned to provide the desired functionality of these using only Web Services technology.

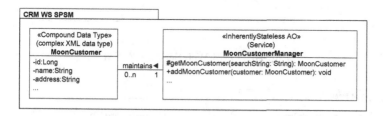

Fig. 15. CRM Web Services SPSM

The SPSM as depicted in Fig. 15 defines the entities used to realize the CRM system in terms of the SPSM metamodel for Web Services. Therefore, the CRM system defines a flat data type that represents the customer data and an object of the type *InherentlyStatelessArchitecturalObject* that manages instances of the customer data type. Note that the abstraction of the *MoonCustomerManager* likewise represents an object type whose instances are observably stateful according to the SPSM Metamodel for Web Services. The methods offered by the *MoonCustomerManager* abstraction and their return values are dependent on each other, while instances of the type maintain the logically internal state (i.e. customer data) using external resources (e.g. using a database).

We present a similar Web Services SPSM for the OMS in the following. The *OrderManager* as presented in Fig. 16 has been realized also as an object of the type *InherentlyStatelessArchitecturalObject* representing an observably stateful abstraction like the CRM manager presented before. It therefore offers multiple dependent operations that all operate on orders. Like in the CRM, SPSM it also contains flat XML data types used for the data exchange of its operations' parameters and return values. These data types are all of the kind *Compound-DataType* and relate to the types *Order, OrderLineItem* and *OrderConfirmation*.

7.2 SPIMs of the Legacy Systems

As a next step of the chosen realization strategy, this section discusses the reverse engineered SPIMs of Moon's legacy systems. We therefore first consider the SPIM of the CRM system.

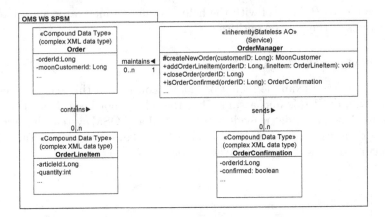

Fig. 16. OMS Web Services SPSM

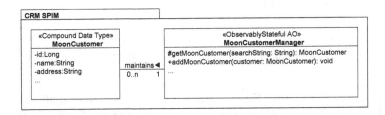

Fig. 17. CRM SPIM

As depicted in Fig. 17, there is not much difference between the SPIM and the SPSM as we decided not to apply any of the patterns at this point. The only difference to the SPSM is the consideration of the *MoonCustomerManager* entity as *Observably Stateful AO* since at the PIM-level the notion of inherent state does not exist.

Finally, we present the SPIM of the OMS system in Fig. 18 that, like the previous SPIM, is directly reverse engineered from the given SPSM. Also, for this model we did not apply any patterns in the step of reverse engineering and the only change is applied to the class *OrderManager* that is also defined as *ObservablyStatefulArchitecturalObject* on the PIM-level.

7.3   CPIM of the Blue-Moon Client Application Scenario

As we discussed in the introduction of this section, the goal of the Mediator CPIM is to provide the client developer an optimized view on the client application scenario's required software entities (i.e. the services) provided by the CRM and the OMS legacy backend systems. As the interface of the Mediator to Blue is defined as a single invocation that receives a purchase order and

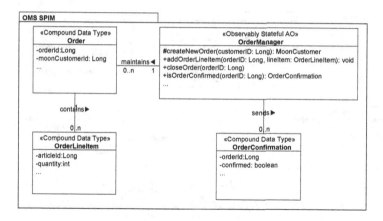

Fig. 18. OMS SPIM

expects a purchase order confirmation as return value, it is the best to de-velop the mediator as a typical client process. We furthermore accommodate the object-oriented metaphor in designing this client process, defining differ-ent kinds of object abstractions for an optimized client-oriented perspective on Moon's provided service landscape as well as an optimized way to develop the mediator client. In Fig. 19, we therefore present the CPIM of the mediator client application scenario as could have been designed by a client developer. The in-formation about Moon's service landscape available for the client developer is given by the specifications (i.e. by the models) of the SPIM and the SPSM. As we have chosen realization strategy (a) we consider the SPIM and its entities in the following. Therefore the client developer's knowledge about the scenario consists of a few given data types and two architectural objects that manage these data types to provide also some rich functionality. In the following we will present the CPIM of the mediator client application scenario and apply some of the presented patterns of Sect. 5.3 for the definition of the CPIM entities.

Since in service-oriented systems compound data types usually represent the *Business Objects* that are managed and manipulated, the focus is first set on these as they can be regarded as first-class citizen domain objects from a client-oriented perspective. This can be considered as the basic appliance of the *Data Type Reification* pattern as introduced in Sect. 5.3. In the CPIM we therefore define the following ephemeral objects which are the *BluePurchaseOrder*, the *MoonPurchaseOrderConfirmation*, the *Order*, the *OrderLiteItem* and finally the *MoonCustomer*. All these ephemeral objects in the CPIM are defined as private since instances of these are exclusively used by one instance of the mediator pro-cess at a time and they are not shared between multiple client instances of the mediator. Three of these ephemeral objects are further classified as *DataObject* as they merely represent the values of the data types and no rich-behaviour is required. These objects are the *BluePurchaseOrder*, the *MoonPurchaseOrder-Confirmation* and finally the *OrderLineItem*.

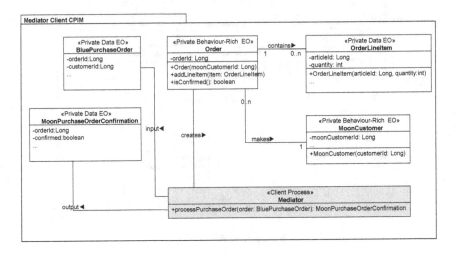

Fig. 19. Mediator CPIM

Considering the functionality of the *OrderManager* in the SPIM, it is obvious at first sight that its offered methods can be seen as the responsibilities of a rich domain object *Order* as would in general be the result of domain analysis and modeling. Thus, we decide to define the *Order* entity in the CPIM as *Behaviour-Rich Object* that has these responsibilities assigned as allocated processes. As the confirmation of an order, which is represented in the OMS SPIM by the *PurchaseOrderConfirmation* data type, can also be regarded as a responsibility of the *Order* object, we can omit the confirmation as a separate object in the CPIM and define an operation assigned to the *Order* object that logically delivers the same information, but allocated to the entity it belongs to.

The same can be done for the domain object *MoonCustomer* of the CRM service that maintains an internal customer id required to carry out further steps in the process. Instead of assigning this responsibility of the service as a usual method we assign it as a special case of an allocated process, that is, within the create method (i.e. the constructor) of this domain object. As internally this requires an external participating resource and cannot be realized with a self-contained entity as *DataObject* is, we regard the *MoonCustomer* as a *Behaviour-RichObject* also.

In summary, to accommodate the object-oriented principles in the CPIM we have encapsulated certain functionality that is provided in a function-oriented way as defined in the SPSMs of the Mediator scenario within rich domain objects (either in the constructors or in additional methods). This ultimately achieves the same goal, but the CPIM abstractions are much more reusable artifacts that are much easier to understand and apply, both, in the process of modeling and in developing based on domain objects with rich behaviour. In the next section we will discuss how the PIM-level abstractions can deliver the expected behaviour at run-time on the PSM-level.

7.4   CPSM Realization

In an earlier section of this paper we stated that we do not expect radical changes of the CPIM abstractions defined in the refinement of these towards abstraction in a CPSM. This is also reflected in the depicted Java CPSM in Fig. 20.

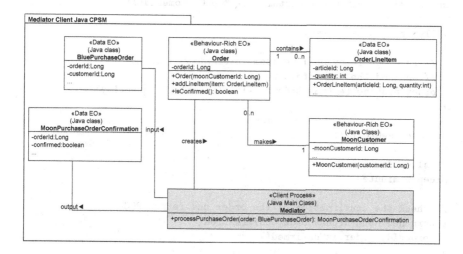

Fig. 20. Mediator Java CPSM

As presented there have been no real changes applied to the entities. However, as we have discussed in Sect. 6.2, the client-oriented properties that have been part of the CPIM have been removed from the single entities. This is done as on the PSM-level it is expected to (automatically or manually) create stubs that already include the behaviour as specified in the CPIM. This significantly reduces the burden of the client developer to deal with these issues. In general, the creation of these stubs can only be carried out (automatically or manually) with the knowledge of all the models (i.e. client- and service-oriented PIMs/PSMs) which are part of the overall application scenario. This is because it is necessary to bridge the gap between the abstractions defined from a client-oriented perspective and the abstractions as provided by the service-oriented perspective.

As an example of how the Mediator client process could be implemented in an object-oriented style using Java according to the defined model and according to the provided platform-specific stubs, we show an appropriate code snippet in pseudo-code notation in Listing 7.1.

This little code snippet shows how the artificial complexity introduced by the interfaces to service-oriented systems on the PSM level can be reduced by client-oriented definitions of the involved entities. As discussed at the beginning of this section, the challenge on the PSM-level is the creation of the platform-specific stubs to be used in programming the client application. Therefore the stubs need to (i) adhere to the characteristics of the entities as specified in the CPIM and (ii)

Listing 7.1.

```
MoonPurchaseOrderConfirmation MediatorClientProcess(PIP3a4PO po)
{
  MoonCustomer customer = new MoonCustomer(po.getCustomerId());

  // Create a new order providing the internal Moon Customer ID
  Order order = new Order(customer.getMoonCustomerId());

  while (po has lineItems)
  {
    // read out relevant line item information from po
    // and add the line items to the order
    order.addLineItem(new OrderLineItem(articleID, quantity));
  }

  order.closeOrder();

  // sleep for 5 minutes
  sleep(5 minutes)

  // check the confirmation of the order and return POC
  return new MoonPurchaseOrderConfirmation(
      orderId, order.isConfirmed()
  );
}
```

furthermore internally need to provide the mapping between the abstraction as specified in the CPIM/CPSM and the abstraction as provided by the SPSM. In a previous paper [1] we have introduced an example that allows these mappings to be carried out automatically, on a one to one basis between the different perspectives. However, as can be recognized in the BlueMoon example in this paper, this is not possible and requires a human, referred to the integrator, who has to manually create these mappings.

In the example of this section there are two entities – the *MoonCustomer* and the *Order* objects – that require the intervention of the human integrator. First we consider the defined abstraction of the *MoonCustomer*. The create operation of this entity is defined as taking the customer ID provided in the *BluePurchaseOrder* object as a parameter. Internally, the responsibility of this method is to call the CRM service of Moon to retrieve the internal Moon customer ID and to initialize the *MoonCustomer* object with the received value. These steps (i.e. the internal invocation of the CRM service) currently have to be added in the *MoonCustomer* stub by the human integrator. However, in the end this stub represents a ready-to-use, reusable artifact that may be used in other client application scenarios (i.e. in other CPIM/CPSMs) also.

The same integration work needs to be applied for the *Order* abstraction as defined in the CPSM. Therefore internally the create operation for the *Order* object needs to invoke one of the OMS services to create a new order. Furthermore the operation to add line items to an order likewise needs to invoke the respective operation of the OMS service and finally the confirmation operation of the defined CPSM *Order* entity needs internally to invoke the confirmed operation of the OMS service. Applying this metaphor the client developer is shielded from function-oriented service invocations and does not need to worry about the introduced complexity of mapping IDs to data structures etc. As with the *Customer* object, this *Order* object represents a reusable artifact that may also be reused in other application scenarios for the development of client applications.

As this example shows, it still requires manual work in some cases to create the stubs on the PSM-level and it cannot be fully automated, but in the future we hope this will be supported by tools that guide the client developer as well as the integrator in the creation of PSM-level artifacts.

8   Related Work

Since Service Oriented Computing (SOC) and Model-Driven Architecture (MDA) are such important paradigms, there has naturally been a great deal of interest in using them together, and over the last decade a number of high profile proposals have been put forward for doing so. These range from life-cycle spanning, industrial scale methods such as SOMA [25] and SOMF [26] to more academic contributions such as [27] and [28]. As a dedicated language for modeling SOAs, the Service-oriented architecture Modeling Language (SoaML) has been standardized by the OMG in [29]. The Service-Oriented Modeling and Architecture (SOMA) approach published by IBM in 2004 [25] was one of the first fully fledged methods targeted at service-oriented architectures. However, it is a very broad spectrum method that covers many more aspects of the development lifecycle than just the modeling and documentation of the service in a service-oriented architecture. The same goes for the Service-Oriented Modeling Framework (SOMF) developed by Michael Bell [26]. It covers everything from the modeling of business goals and processes to enterprise architectures.

Probably the method most focused on supporting the UML-based modeling of service-oriented architectures is the method of Piccinelli and Skene [28], which specializes in the modeling of Electronic Service Systems (ESSs). This uses a mixture of metamodels and profiles (the ESS profile) to support two views of services – one (the capability view) showing how services are composed from capabilities and the other (the implementation view) showing how business assets are assigned to capability roles to make services concrete. However, both views are highly abstract and platform-independent, and provide no support for modeling SOA's realized in a particular execution platform. The method of Lopez-Sanz et al. [27] also aims to support the modeling of service-oriented software architectures using MDA principles. It does this within the context of the MIDAS model-driven methodological framework by defining a single metamodel that consists of typical concepts from service-oriented computing.

Like [28], however, it also focuses exclusively at the PIM level only from a single perspective.

The big difference between these approaches and the unified conceptual model presented in this paper is that they focus on modeling services at only one level of abstraction (PIM level) and usually only from one perspective. Where more than one perspective is supported (e.g. in [28]) these are not focused on separating the client developer's and the service provider's perspectives. The approach presented in this paper is unique in:

- highlighting the four possible combinations of the PIM versus PSM perspectives and the client-oriented versus service-oriented perspective as the four most important viewpoints from which to visualize SOAs,
- identifying a core set of concepts common to all these viewpoints, and
- specializing these core concepts within four distinct metamodels, optimized for the modeling of services and client artifacts from the point of view of each distinct stakeholder.

An important specialty of the patterns discussed in this article is that they are transformational. They deal with relationships or morphisms between kinds of types and objects. This is a crucial difference to well-established OO patterns found in literature. Consider, as an example, the *Ephemeral Object Manager* pattern introduced and discussed in Sect. 5.3. It describes the transformation of an ephemeral object into an architectural object by the introduction of object managing functionality. The resulting object manager resembles the proven factory/finder pattern from the CORBA services framework [30], which itself is a pattern that evolved from the combination of other proven OO patterns, i.e., the proxy and singleton pattern [32]. However, for our purposes, the resulting object manager structure is not sufficient by itself, it is the ideas wrapped up in the transformation that represent the more important pattern for our purposes.

Many of the concepts introduced in this article may seem similar to the CORBA [5] and J2EE [6] frameworks, because these also tried to facilitate distributed computing in terms of object concepts. However there are two main differences to our approach. The first is that we focus on client-side concerns, and on reducing the artificial complexity experienced by client developers when accessing service infrastructures. Second, CORBA and J2EE both focus on the forward-driven construction of distributed systems (from requirements to components) whereas we place equal, if not more, emphasis on the reuse of existing server-side assets (from components to requirements).

Our approach is related to, and builds on, existing conceptual models for distributed computing such as RM-ODP and COSMO [31] as well as on general component models such as CORBA Component Model [5] and the Service Component Architecture (SCA) [12]. The main difference is that our approach incorporates different views that support the creation of models that are customized for the different stakeholders.

9   Conclusion

In this paper we have presented a unified conceptual framework for describing the components of service-oriented computing systems and client development at two key levels of abstraction (platform- independent and platform-specific) and from the perspective of two key roles or stakeholders (service provider and client developer). By separating concerns in this way it is possible to support distinct views of a system customized for different stakeholders and optimized for their specific needs. In particular, client developers can be provided with an object-oriented viewpoint of the abstractions involved in a particular business process or application, and can thus be relieved of the burden of writing code (or process specifications) to interact with services at the level of abstraction that has been optimized for maximum interoperability rather than for usability by clients.

The contribution of this approach lies not only in the shape and structure of the overall modeling framework but also in the separation of concerns and identification of common abstractions that is reflected in the contents of the various metamodels. This is not only the prerequisite for integrating most of the current distributed computing and client programming technologies into the framework in the form of customized platform-specific specializations of the common abstractions, it also facilitates the definition of the basic realization patterns which can be used to support the core client-oriented abstractions on top of all compliant server-side technologies.

The insights into tiers of services and their usage by clients presented in the framework can be exploited immediately in projects that deal with the design of distributed systems based on today's technologies, which are usually technologically heterogeneous based on a mix of object-oriented, service-oriented and scripting technology. The framework can be also exploited in the design of new lightweight mapping tools that combine the benefits of the interface definition compilers from heavyweight distributed object computing technology with the advantages of today's lightweight service-oriented technology and architecture.

Currently we are expanding the number of concrete implementation platforms incorporated into the framework (as platform-specific models) and are applying the approach to a wider number of case studies.

References

1. Atkinson, C., Bostan, P., Draheim, D.: Towards a Unified Conceptual Framework for Service-Oriented Computing. In: Proceedings of 3M4SE 2011 - Proceedings of 2nd International Workshop on Models and Model-driven Methods for Service Engineering. IEEE Press (September 2011)
2. Atkinson, C., Bostan, P.: Towards a Client-Oriented Model of Types and States in Service-Oriented Development. In: Thirteenth IEEE International EDOC Conference, EDOC 2009, Auckland, New Zealand (August/September 2009)
3. Gray, J., Reuter, A.: Transaction Processing: Concepts and Techniques. Morgan Kaufmann (1993)

4. Bernstein, P.A.: Middleware: a Model for Distributed System Services. Communications of the ACM 39(2), 86–98 (1996)
5. Object Management Group. CORBA Component Model Specification, OMG Available Specification, version 4.0, formal/06-04-01 dtc/06-02-01 (April 2006)
6. Kassem, N., The Enterprise Team: Designing Enterprise Applications with the Java 2 Platform, Enterprise edn. Sun Microsystems (2000)
7. Fielding, R.T., Taylor, R.N.: Principled Design of the Modern Web Architecture. ACM Transactions on Internet Technology 2(2), 115–150 (2002)
8. Wang, L., Von Laszewski, G., Younge, A., He, X., Kunze, M., Tao, J., Fu, C.: Cloud computing: A Perspective study. New Generation Computing 28(2), 137–146 (2010) ISSN 0288-3635
9. Hess, A., Humm, B., Voß, M., Engels, G.: Structuring Software Cities A Multidimensional Approach. In: EDOC 2007, pp. 122–129 (2007)
10. Haas, L.: Building an Information Infrastructure for Enterprise Applications. In: Draheim, D., Weber, G. (eds.) TEAA 2005. LNCS, vol. 3888, p. 1. Springer, Heidelberg (2006)
11. Draheim, D.: The Service-Oriented Metaphor Deciphered. Journal of Computing Science and Engineering 4(4) (December 2010); Lee, I., Park, J.C., Song, I.
12. Open Service Oriented Architecture. SCA Assembly Model V1.00 (March 2007)
13. Brown, A., Johnston, S., Kelly, K.: Using Service-Oriented Architecture and Component-Based Development to Build Web Service Applications. Rational Software Corporation, Santa Clara (2002)
14. Carr, N.G.: The Big Switch – Rewiring the World, from Edison to Google. W. W. Norton & Company (2008)
15. Atkinson, C., Draheim, D.: Cloud Aided-Software Engineering - Evolving Viable Software Systems through a Web of Views. In: Mahmood, Z., Saeed, S. (eds.) Software Engineering Frameworks for Cloud Computing Paradigm. Springer (2013)
16. Schulte, R.W., Natis, Y.V.: Service Oriented Architectures, Part 1. Gartner Research ID Number SPA-401-068. Gartner (1996)
17. Schulte, R.W.: Service Oriented Architectures, Part 2. Gartner Research ID Number SPA-401-069. Gartner (1996)
18. Emmelhainz, M.A.: EDI: A Total Management Guide. Van Nostrand Reinhold (1993)
19. Holley, K., Palistrant, J., Graham, S.: Effective SOA Governance. IBM White Paper, IBM Corporation (March 2006)
20. Hummel, O., Atkinson, C.: Supporting Agile Reuse Through Extreme Harvesting. In: Concas, G., Damiani, E., Scotto, M., Succi, G. (eds.) XP 2007. LNCS, vol. 4536, pp. 28–37. Springer, Heidelberg (2007)
21. Atkinson, C., Bostan, P., Hummel, O., Stoll, D.: A Practical Approach to Web Service Discovery and Retrieval. In: Proceedings of ICWS 2007 - the 5th IEEE International Conference on Web Services. IEEE Press (2007)
22. Ousterhout, J.K.: Scripting: Higher-Level Programming for the 21st Century. Computer 31(3), 23–30 (1998)
23. Soley, R.: Model Driven Architecture, white paper formal/02-04-03, draft, 3.2. Object Management Group (November 2003)
24. Lutteroth, C., Draheim, D., Weber, G.: A Type System for Reflective Program Generators. Science of Computer Programming 76(5) (May 2011)
25. Arsanjan, A.: Service-Oriented Modeling & Architecture. IBM Online article (November 09, 2004)
26. Bell, M.: Introduction to Service-Oriented Modeling. In: Service-Oriented Modeling: Service Analysis, Design, and Architecture. Wiley & Sons (2008)

27. Lopez-Sanz, M., Acura, C.J., Cuesta, C.E., Marcos, E.: Defining Service-Oriented Software Architecture Models for a MDA-based Development Process. In: 7th Working IEEE/IFIP Conference on Software Architecture (2008)
28. Picinelli, G., Skene, J.: Service-oriented Computing and Model Driven Architecture. In: Service-Oriented Software Systems Engineering: Challenges and Practices. Idea Group Inc. (2005)
29. OMG. Service oriented architecture Modeling Language (SoaML) SoaML OMG Specification (2009), http://www.omg.org/spec/SoaML/
30. Object Management Group. CORBAservices: Common Object Services Specification (March 1995)
31. Quartel, D., et al.: COSMO: A conceptual framework for service modelling and refinement. Information Systems Frontiers 9(2-3), 225–244 (2007), doi:10.1007/s10796-007-9034-7
32. Coplien, J.O., Schmidt, D.C. (eds.): Pattern Languages of Program Design. Addison-Wesley (1995)
33. Semantic Web Services Challenge 2006 (2006), http://sws-challenge.org

Author Index